SEASONAL CARBON CYCLING IN THE SARGASSO SEA NEAR BERMUDA

Seasonal Carbon Cycling in the Sargasso Sea near Bermuda

Nicolas Gruber and Charles D. Keeling

UNIVERSITY OF CALIFORNIA PRESS
Berkeley • Los Angeles • London

BULLETIN OF THE SCRIPPS INSTITUTION OF OCEANOGRAPHY
OF THE UNIVERSITY OF CALIFORNIA, SAN DIEGO
LA JOLLA, CALIFORNIA

Advisory Editors: Charles S. Cox, Gerald L. Kooyman,
Richard H. Rosenblatt (Chairman)

Volume 30

Approved for publication August 1997

UNIVERSITY OF CALIFORNIA PRESS
BERKELEY AND LOS ANGELES, CALIFORNIA

UNIVERSITY OF CALIFORNIA PRESS, LTD.
LONDON, ENGLAND

Library of Congress Cataloging-in-Publication Data

Gruber, Nicolas, 1968–.
    Seasonal carbon cycling in the Sargasso Sea near Bermuda / Nicolas
Gruber and Charles D. Keeling.
       p.   cm. — (Bulletin of the Scripps Institution of Oceanography,
University of California, San Diego; v. 30)
    Includes bibliographical references.
    ISBN 0-520-09833-1 (paper: alk. paper)
    1. Seawater—Analysis. 2. Carbon—Analysis.   3. Carbon cycle
(Biogeochemistry)   I. Keeling, Charles D., 1928–   . II. Title.
III. Series.
GC117.C37G78   1999
551.46'462—dc21   99-32243
                                          CIP

CONTENTS

ABSTRACT

By observing simultaneous changes in the concentration and $^{13}C/^{12}C$ ratio of dissolved inorganic carbon (DIC) in seawater over the annual carbon cycle it is possible to distinguish physical and biological processes that govern this cycle in near-surface waters. We report an analysis of monthly data from 1983 through 1989 at Station S in the surface mixed layer of the Sargasso Sea near Bermuda. The analysis also makes use of contemporary measurements of alkalinity, the calculated CO_2 partial pressure in seawater, and measurements of atmospheric CO_2 concentration and $^{13}C/^{12}C$ ratio from which the net flux of CO_2 across the air-sea boundary was calculated as a function of wind-speed for each isotopic species. Limited additional chemical measurements in deeper water and physical oceanographic data afforded estimates of the seasonally varying upward flux of each isotopic species of DIC into the mixed layer by vertical diffusion and entrainment. Net community production (the net transfer of carbon from inorganic to organic pools) was estimated for the mixed layer from the observed change in the $^{13}C/^{12}C$ ratio of DIC after corrections were applied for air-sea exchange and vertical transport and using the fact of a strong fractionation of ^{13}C during the photosynthetic uptake of DIC. As a second method, and to check our first approach, net community production was also calculated simply by difference from the observed change in DIC, after also taking into account air-sea exchange and vertical transport. Thus, both these estimates are obtained without reference to nutrient fluxes or concentrations of organic matter. We deduce that 11 gC m^{-2} yr^{-1} of DIC is lost on average as a result of a net community production of 19 gC m^{-2} during shoaling of the mixed layer in spring and summer and a reversed net transfer from organic to inorganic of 8 gC m^{-2} during deepening. Air-sea exchange adds 21 gC m^{-2} yr^{-1} of CO_2 to the mixed layer and vertical transport a further 18 gC m^{-2} yr^{-1} of DIC. The sum of all fluxes is thus a net apparent gain of 28 gC m^{-2}, which is nearly balanced by an apparent loss term associated with the variable mixed-layer depth. We have assessed the uncertainties in our calculations by sensitivity tests. Our estimate of the annual air-sea exchange lies between previously published estimates based on atmospheric modeling studies. When, on the basis of limited observational evidence we include approximately 7 gC m^{-2} yr^{-1} of net community production below the mixed layer, the resulting tentative estimate of net community production in the entire euphotic layer is 18 gC m^{-2} yr^{-1}. This estimate is comparable to those of net export of organic matter based on sediment traps and on water-column inventories, but considerably lower than estimates of euphotic new production based on measurements of oxygen and of noble gases and on other modeling studies for Station S.

ACKNOWLEDGMENTS

We are grateful to the staff of the Bermuda Biological Station for Research, Inc.: to the director, Anthony Knap, and to Timothy Jickels, Rachael Sherriff-Dow, and Anthony Michaels for their continuing assistance in providing water samples and for their many helpful discussions regarding both the field work and the interpretation of the data. We particularly thank Peter Brewer of the Woods Hole Oceanographic Institution for persuading us to undertake this study. We also thank Robert Bacastow, Peter Guenther, Tom Hayward, Ralph Keeling, Timothy Lueker, Pearn Niiler, Mark Ohman, Stephen Piper, and Russ Davis of the Scripps Institution of Oceanography, Olivier Marchal of Centre des Faibles Radioactivités, France, Thomas Stocker of the University of Bern, and Jorge Sarmiento of Princeton University for valuable discussions during preparation of this article. Timothy Lueker prepared many of the data for use in modeling. Elisabeth Stewart assisted in programming and preparing graphic material and deserves special thanks for her very careful checking of the computations and programming logic. We are indebted to Robert Bacastow for calculating gradients in DIC and $^{13}\delta$ using a three-dimensional transport model, and we thank Ernst Maier-Reimer of the Max Planck Institute for Meteorology in Hamburg, Germany, for kindly making this model available to us. We thank Mr. and Mrs. Fritz Gruber for their generous financial support for the first author during his stay at the Scripps Institution of Oceanography. We also acknowledge the financial support of Prof. Thomas Stocker of the University of Bern for the first author during later revisions of this article. Financial support was from the National Science Foundation, Grants OCE-82-08475, ATM-85-16939, ATM-88-19398, and ATM-91-21986; the U.S. Department of Energy via Grant FG03-90ER-60982; and the Electric Power Research Institute, Contract RP8011-20.

1. INTRODUCTION

Physical and biogeochemical processes act together in controlling the carbon balance in the upper ocean. The international oceanic sciences community recently made remarkable progress toward establishing details of this marine carbon cycle based on results from the ongoing Joint Global Ocean Flux Study time series stations in the subtropical North Atlantic (Bermuda Atlantic Time-series Study, near Bermuda) and the subtropical North Pacific (Hawaii Ocean Time-Series, near Hawaii) (Michaels et al. 1994a, Michaels et al. 1994b, Winn et al. 1994, Karl and Lucas 1996, Michaels and Knap 1996). However, underlying problems, such as the extent of biological production and of nutrient supply to near-surface waters, are not yet solved for major oligotrophic oceanic regions in mid- and low-latitudes (Lewis et al. 1986, Jenkins 1988, Altabet 1989a, Villareal, Altabet, and Culver-Rymsza 1993, Hayward 1994, Michaels et al. 1994a, Bates, Michaels, and Knap 1996). Further uncertainties include the magnitude of the contributions of air-sea exchange and transport processes to the observed temporal and spatial inorganic carbon distribution (Michaels et al. 1994b, Toggweiler 1994, Bates, Michaels, and Knap 1996).

We address these issues here by interpreting a series of inorganic carbon measurements made at Station S near Bermuda over a six-year period from 1983 to 1989. These measurements, mainly confined to near-surface waters, yield a representation of the seasonal cycle of carbon in the mixed layer of the western Sargasso Sea in greater detail than previously possible. Both dissolved inorganic carbon (DIC) and titration alkalinity (TA) were determined, thus permitting dissolved CO_2, and hence the CO_2 partial pressure (pCO_{2oc}), to be calculated from thermodynamic relations. In addition the stable isotopic ratio, $^{13}C/^{12}C$, of DIC was measured. Because carbon isotopes fractionate during photosynthesis, variations in this ratio in DIC are sensitive to the interconversion of DIC and organic carbon. Thus $^{13}C/^{12}C$ data permit an estimate to be made of the net community production of organic matter in the mixed layer without recourse to less direct approaches, which depend on the association of the oceanic carbon cycle with the seasonal cycling of nutrients and dissolved oxygen.

To establish the seasonal cycling of carbon from our data we have developed a spatial model focusing on the budget of inorganic carbon in the mixed layer, which is regarded as a vertically homogeneous compartment in contact with the atmosphere above and the waters of the thermocline below. We consider the rates of carbon species change in this compartment, and the inferred fluxes of carbon in and out of this compartment. We take into account physical variables, including temperature, salinity, and wind speed, to estimate these fluxes. To test our results we make use of sensitivity tests and a comparison with predictions for the Sargasso Sea provided by a global three-dimensional carbon transport model, which takes into account large-scale lateral transport not directly detectable in our study at a single geographic point. Because our observational data for the vertical gradient in $^{13}C/^{12}C$ below the mixed layer cannot be proven to be representative, we also use this model as an aid in estimating this gradient.

As we will demonstrate, $^{13}C/^{12}C$ data offer a promising approach to characterizing the oceanic carbon cycle. This approach would have produced even more valuable insight if monthly measurements could have been made throughout the upper water column, but such

1

an expanded program was not feasible. Here we explore the value of including such isotopic data generally in the study of the oceanic carbon cycle.

Even though this document should be regarded as only a preliminary assessment of the seasonal cycle of carbon, it may provide a useful complement to studies focused more directly on organic processes, rather than on inorganic carbon. The inorganic carbon cycle should be fully characterized to establish the relative importance of oceanic and terrestrial processes in regulating the abundance of carbon in the principal reservoirs of the global carbon cycle; it should also be better understood as a complement to any complete understanding of organic processes in the ocean.

This bulletin is organized as follows. In the next two sections we review knowledge of the carbon cycle in the surface oceans. We include a short general discussion on the use of concurrent measurements of DIC and its isotopic ratio to constrain carbon budgets. Readers familiar with these topics may skip these two sections and go directly to Section 4, where we describe the seasonal observations obtained at Station S. These observations form the input to a seasonal box model, explained in Sections 5 and 6. Then in Sections 7 and 8, we discuss our findings and compare them with previous estimates of biological production near Bermuda and with predictions from a three-dimensional carbon transport model as noted earlier. We present our conclusions in Section 9. Details of the model formulation and its application are placed in a series of appendices.

2. PROCESSES CONTROLLING THE CARBON BALANCE IN THE UPPER OCEAN

2.1 INTRODUCTION

We begin by discussing the processes of both biological and physicochemical origin that influence the balance of DIC in the upper ocean with special emphasis on their effects on the stable isotopic ratio of DIC, expressed by the reduced ratio:

$$^{13}\delta = \frac{r - r_s}{r} \, , \tag{1}$$

where r denotes the $^{13}C/^{12}C$ ratio of the sample and r_s the $^{13}C/^{12}C$ ratio of the belemnite carbonate standard, PDB. Because the values of $^{13}\delta$ are small, they are expressed in per mil (‰). Here and subsequently, DIC denotes the analytical sum of all inorganic dissolved carbon species, i.e. $[CO_2]_{aq}$ + $[H_2CO_3]$ + $[HCO_3^-]$ + $[CO_3^{2-}]$ and all chemical ligands of these species with other anions and cations (SCOPE 1981).

2.2 BIOLOGICAL PROCESSES

Within the ocean's near-surface layer, organic matter is formed by photosynthesis of phytoplankton, whereby nutrients are taken up and oxygen is released. This sunlit "euphotic" layer, which generally extends to the 1% light level (Valiela 1984, 40, Morel 1988), constitutes approximately the upper 100 m of the water column near Bermuda (Menzel and Ryther 1960, 365, Marra et al. 1992).

The chemical formulation of photosynthesis and respiration is often summarized by the following simplified equation (Redfield, Ketchum, and Richards 1963):

$$106 \ CO_2 + 16 \ NO_3^- + HPO_4^{2-} + 122 \ H_2O + 18 \ H^+ =$$
$$(C_{106}H_{263}O_{110}N_{16}P) + 138 \ O_2, \tag{2}$$

where the expression in the bracket is an empirical formula for organic matter. The stoichiometric ratio of carbon uptake to oxygen release during photosynthesis (O_2/CO_2), often called photosynthetic quotient (PQ), is about 1.3 according to this formula. This photosynthetic quotient is not constant, however, and may vary between 1.0 and 1.4 (Williams and Robertson 1991) depending on the source of nitrogen (Laws 1991) and other processes (e.g. photorespiration [Burris 1981]).

In the euphotic layer during photosynthesis, the lighter isotope, ^{12}C, is preferentially taken up compared with the heavier isotope ^{13}C. This results in a low $^{13}\delta$ value for the organic carbon in phytoplankton (between -20 and -30‰ [Degens 1969, Goericke and Fry 1994]) compared to surface water DIC (0 to 2‰ [Kroopnick 1985]). The photosynthezising organisms return a fraction of the fixed organic carbon to CO_2 by autotrophic respiration. Grazing of the phytoplankton by zooplankton and subsequent heterotrophic respiration define another pathway from organic matter back to DIC. Upon the death of these planktonic organisms, the resulting detrital organic pool is remineralized to DIC mainly by heterotrophic bacteria. Additional dissolved organic carbon, released to the water during growth and feeding, is similarly remineralized (Baines and Pace 1991, Longhurst 1991). Little or no isotopic fractionation occurs during the processes of grazing, respiration, and remineralization (Deuser, Degens, and Guillard 1968, 657). Even in the case of the respiration by phytoplankton, in which $^{12}CO_2$ may be preferentially released over $^{13}CO_2$ (Degens et al. 1968, 8), the fractionation effect is believed to be small (O'Leary 1981, 560).

Part of the organic matter formed by photosynthesis is exported out of the euphotic layer via sinking of large particles (Martin et al. 1987). Some of the remainder is exported by diffusion and advection of dissolved organic matter (Bacastow and Maier-Reimer 1991, Najjar, Sarmiento, and Toggweiler 1992, Carlson, Ducklow, and Michaels 1994) and by vertical migration of zooplankton (Longhurst 1991). In the subsurface layers organic matter from the mixed layer is subsequently remineralized again mainly by heterotrophic organisms.

To denote aspects of the production of organic matter and its subsequent cycling through the euphotic food web layer, we adopt the carbon-based terminology of Williams (1993). *Gross primary production*, P_g, denotes the organic carbon produced by the chemical reduction of carbon as a consequence of the photosynthetic process during some specific period of time. *Net primary production*, P_{np}, denotes gross primary production minus the losses in carbon caused by autotrophic respiration, R_a, during the same time period, i.e.;

$$P_{np} = P_g - R_a . \tag{2a}$$

Finally, *net community production*, P_{nc}, denotes gross primary production minus all losses in carbon from respiration, including heterotrophic respiration, R_h, i.e.;

$$P_{nc} = P_g - R_a - R_h. \tag{2b}$$

In pelagic ecosystems, net primary production is further distinguished as *regenerative* when the production is supported by nutrients derived locally within the euphotic layer, or *new*, if the nutrients are supplied from outside (Dugdale and Goering 1967). Eppley and Peterson (1979) assumed that new production is supplied by nitrate or fixed atmospheric molecular nitrogen, whereas regenerated production is based on reduced nitrogen compounds, mainly ammonia derived from the excretion of animals and from the metabolism of heterotrophic microorganisms. The ratio of new production to net primary production we will call the "f-ratio" following Eppley and Peterson (1979). Assuming a quasi steady state over the annual cycle, new production can be equated both to net community production (Platt et al. 1989, Laws 1991) and to the export of organic matter out of this layer (Martin et al. 1987).

Typically, estimates of biological production have been obtained by following, in vitro, the uptake of radioactively labeled carbon or the production of oxygen. Two problems with this technique arise. First, the effect of bottle confinement on the phytoplankton population is uncertain (Bender et al. 1987) and second, obtaining seasonal production estimates is difficult because of possible undetected intermittent production (Jenkins and Goldman 1985, Wiggert, Dickey, and Granata 1994). Attempts to overcome the limitations of in vitro determinations have been made by comparing results with observations of mesoscale changes in dissolved oxygen (Jenkins and Goldman 1985, Musgrave, Chou, and Jenkins 1988, Thomas, Garcon, and Minster 1990, Thomas et al. 1993, Emerson et al. 1993) and in dissolved inorganic carbon (DIC) (Oudot 1989, Karl, Tilbrook, and Tien 1991, Chipman,

Marra, and Takahashi 1993, Robertson and Watson 1993, Bates, Michaels, and Knap 1996). In comparison to techniques in vitro, which yield estimates of gross or net primary production (Williams et al. 1983), this mesoscale approach allows estimates of net community production to be made (Robertson and Watson 1993).

The study of variations in concentration of dissolved oxygen is an imperfect substitute for estimating biological production because this estimate is heavily dependent on the photosynthetic quotient chosen for the conversion from units of oxygen to units of carbon. This limitation does not exist in studies in which observed mesoscale changes of DIC are used to make estimates of production. But usually, in these latter studies, net community production has been estimated by the difference from the observed changes after corrections have been applied for gas exchange at the air-sea boundary, vertical diffusion, and biogenic formation of carbonate particles. The uncertainty in estimation of net community production is, therefore, directly proportional to the combined uncertainties of the other processes considered. An independent check on the consistency of the results would be desirable to reduce this uncertainty, but until now has not been provided.

Sediment traps have been used extensively to infer the export of particulate organic matter from the overlying water (e.g. Martin et al. 1987, Altabet 1989a,b; Karl, Tilbrook, and Tien 1991, Michaels and Knap 1996). If steady-state conditions prevail, the export of total organic matter should be equal to net community production (Laws 1991). However, interpretation of sediment trap measurements is difficult because of sampling problems associated with hydrodynamic biases, interference by swimmers, and loss of dissolved organic carbon (DOC) from the high density solution used to preserve the particulate organic carbon (Karl and Knauer 1989, Michaels et al. 1994b, Buesseler et al. 1994). Furthermore, sediment traps measure only the export of particulate organic carbon (POC), not taking into account the possible export of organic carbon by vertical migration of zooplankton (Longhurst 1991) or by diffusion and advection of dissolved organic carbon (Najjar, Sarmiento, and Toggweiler 1992, Carlson, Ducklow, and Michaels 1994).

Biological removal or regeneration of carbon can occur, however, as a result of photosynthesis, respiration, and remineralization, and as a result of calcium carbonate uptake by producing calcitic or aragonitic organism hard parts. Although this latter process has been shown to contribute significantly to the surface carbon cycle during the spring bloom in the northeast Atlantic (Robertson et al. 1994), it is generally unimportant in the Sargasso Sea near Bermuda (Brown and Yoder 1994, Bates, Michaels, and Knap 1996).

2.3 AIR-SEA EXCHANGE

Gaseous CO_2 is transferred across the air-sea interface whenever the pressure of atmospheric CO_2 above the surface differs from the gas pressure exerted by dissolved CO_2 in the water at the sea surface. Because of the dependency of the solubility of CO_2 in seawater on temperature and salt content (Weiss 1974), this latter "partial pressure," as it is called, reflects not only changes in the dissolved carbon in the water, but also changes in temperature and salinity. A kinetic fractionation occurs during air-sea gas exchange involving the diffusion of CO_2 across the air-sea boundary layer. This kinetic effect is imposed on a temperature-dependent equilibrium fractionation between the gas and liquid phases (Heimann and Keeling 1989). The latter fractionation process promotes lower $^{13}\delta$ values for atmospheric CO_2 in comparison to the $^{13}\delta$ of DIC. The $^{13}\delta$ difference between air and sea is further influenced by the Suess effect (cf. Heimann and Keeling [1989, 264]) in which the release of CO_2 from fossil fuel, with a $^{13}\delta$ of around -27‰ (Tans 1981, 128) has led to a drop of $^{13}\delta$ of atmospheric CO_2 over the last century from -6.4‰ (Friedli et al. 1986) to around -7.7‰ (Keeling et al. 1989, 186). Therefore, CO_2 entering the ocean from the atmosphere has a significantly more negative isotopic signature than in preindustrial times (Quay, Tilbrook, and Wong 1992, Tans, Berry, and Keeling 1993, Bacastow et al. 1996).

2.4 MIXED-LAYER DYNAMICS AND VERTICAL TRANSPORT

Close to the surface of the ocean, convective processes and input of turbulent kinetic energy by wind stress sustain a mixed layer with almost uniform vertical profiles of temperature and chemical properties. The depth of this mixed layer varies on time scales from days to seasons according to the variability in the driving processes; it rarely corresponds to the depth of the euphotic layer. Therefore the concentration of DIC and other chemical properties is likely to vary vertically in the euphotic layer, whenever it is deeper than the mixed layer. During times of deepening, the mixed layer incorporates water from below with different physical and chemical properties, a process we call entrainment (Phillips 1977). Consequently, heat, momentum, salt, and other chemical constituents are transferred across the lower boundary of the mixed layer at a rate proportional to the rate of deepening, and to the gradient just below this boundary. In contrast, during shoaling of the mixed-layer depth no such transport occurs.

Vertical turbulent diffusion also transports heat and chemical constituents across the lower boundary. This turbulent flux is classically formulated by analogy to molecular diffusion. The two processes of entrainment and vertical diffusion are difficult to distinguish experimentally, although they are clearly separated conceptually. The fluxes of DIC caused by entrainment and vertical diffusion are not accompanied by isotopic fractionation because both processes, being turbulent, do not discriminate at the molecular scale.

Viewing the ocean on a regional or global scale, the vertical gradient of DIC below the mixed layer is affected by remineralization of the exported organic matter (the soft tissue pump), the buildup of anthropogenic CO_2 (the Suess effect), and the temperature-dependent solubility of CO_2 (the solubility pump). Although not further addressed in this study, it is also affected by the formation and dissolution of carbonate particles (carbonate pump) (Volk and Hoffert 1985). As a result, the concentration of DIC in the subsurface ocean usually increases with depth (Takahashi, Broecker, and Bainbridge 1981), thereby promoting an upward flux by eddy diffusion and by entrainment into the mixed layer. The organic and solubility pumps produce a decrease in $^{13}C/^{12}C$ ratio with depth. This decrease is mainly a result of respiration and of remineralization of organic matter with a lower $^{13}C/^{12}C$ ratio than DIC at the sea surface. The buildup of fossil fuel CO_2 with a lower $^{13}C/^{12}C$ ratio (Suess effect) and temperature-dependent fractionation, promoting a generally higher ratio below the mixed layer, partially counteract the gradients produced by respiration and remineralization. Thus entrainment and turbulent diffusion transport DIC into the mixed layer with a lesser isotopic signature than that of organic carbon, even though these dynamic processes, by themselves, do not fractionate the isotopes of carbon.

2.5 LATERAL TRANSPORT

Lateral transport, although more difficult to detect than air-sea exchange and vertical transport, also is likely to contribute significantly to the local budget of DIC (Michaels et al. 1994b, Toggweiler 1994). This can be demonstrated by the finding that broad regions of the oceans are persistent sources or sinks of atmospheric CO_2 not accounted for by the buildup of CO_2 from fossil fuel combustion (Tans, Fung, and Takahashi 1990). These regions can only be in balance locally when a net gain or loss of CO_2 by air-sea exchange is compensated for with lateral transport by ocean currents (Winn et al. 1994). As in the case of entrainment and vertical diffusion, no isotopic fractionation is involved, but gradients produced by other processes cause isotopic effects.

2.6 THE SEASONAL CYCLE NEAR BERMUDA

In the euphotic layer near Bermuda a strong seasonal cycle develops every year as a result of deep convective mixing of the waters in late winter followed by shoaling of the mixed layer and the development of a strong summer thermocline (Menzel and Ryther 1960, Jenkins and Goldman 1985, 468; Michaels et al. 1994a). The cycle can be thought to begin each spring when influxes of DIC and nitrate, by convective and wind-driven mixing, produce maximum concentrations of these chemicals near the ocean surface. The availability of nitrate, coupled with a shoaling mixed layer caused by increased irradiance and reduced wind stress, leads to favorable conditions for phytoplankton growth. This results in a moderate diatom-dominated spring bloom (Menzel and Rhyther 1960, Siegel et al. 1990, Malone, Pike, and Conley 1993, Michaels et al. 1994a). Usually nitrate is exhausted after a time and further supplies of nutrients are effectively cut off by a strong summer thermocline. The ecosystem then is presumed to run predominantly on regenerated nitrogen (Malone, Pike, and Conley 1993). Thus changes in magnitude and type of biological production are driven mainly by variability in vertical transport (Siegel et al. 1990). Near the end of summer, surface cooling and stronger winds cause the mixed layer to deepen more or less gradually through the autumn and winter, until the maximum depth is reached in late winter. Because this maximum depth and the coupled vertical transport of nitrate vary considerably from year to year, interannual variability of biological production near Bermuda is common (Menzel and Ryther 1961, Michaels et al. 1994a, Michaels and Knap 1996).

3. CONSTRAINING CARBON BUDGETS BY CONCURRENT MEASUREMENTS OF DIC AND $^{13}\delta$

To demonstrate how concurrent measurements of DIC and its isotopic ratio $^{13}\delta$ constrain carbon budgets we now develop the following budget equations to be used in our later analysis.

Let ΔDIC and $\Delta^{13}\delta$ denote the changes in DIC and $^{13}\delta$, respectively, in a constant volume of water over a time-interval Δt, which begins at t-1 and ends at t. Expressing these changes as sums of changes, ΔDIC_i and $\Delta^{13}\delta_i$, owing to j separately identified processes $i=1\ldots j$:

$$\Delta DIC \equiv DIC_t - DIC_{t-1} = \sum_{i=1}^{j} \Delta DIC_i, \qquad (3)$$

$$\Delta^{13}\delta \equiv {}^{13}\delta_t - {}^{13}\delta_{t-1}. \qquad (4)$$

To connect the overall isotopic change, $\Delta^{13}\delta$ with the individual isotopic changes, $\Delta^{13}\delta_i$, we note that when samples or pools of carbon are mixed, the product of $^{13}\delta$ and carbon mass is nearly an additive function (Mook et al. 1983). More specifically, if $^{13}\delta_1$, $^{13}\delta_2$, $\ldots$, denote the reduced isotopic ratios of a series of carbon pools, N_i, containing N_1, N_2, $\ldots$, moles of carbon respectively, then the mixture obtained by combining the pools has very nearly the $^{13}\delta$ value

$$^{13}\delta_M = \frac{N_1{}^{13}\delta_1 + N_2{}^{13}\delta_2 + \ldots,}{N_1 + N_2 + \ldots,} = \frac{\Sigma N_i{}^{13}\delta_i}{\Sigma N_i}. \qquad (5)$$

Applying this additivity rule to our time-step calculation yields the additional equation:

$$DIC_{t-1}{}^{13}\delta_{t-1} + \Sigma \Delta DIC_i{}^{13}\delta_i = (DIC_{t-1} + \Delta DIC)({}^{13}\delta_{t-1} + \Delta^{13}\delta), \qquad (6)$$

where ΣN_i is equal to $(DIC_{t-1} + \Delta DIC)$, and $^{13}\delta_M$ is equal to $(\Delta^{13}\delta_{t-1} + \Delta^{13}\delta)$. Provided that the changes ΔDIC and $\Delta^{13}\delta$ have been measured, it is possible from equations (3) and (4) to compute the changes in concentration and isotopic ratio of any one process i given independent knowledge of the changes owing to the other j - 1 processes. Equation (6) provides yet another constraint, which is the basis for an independent estimate of biological exchange.

4. SEASONAL OBSERVATIONS

The data utilized in this seasonal study are derived from water samples collected at Station S near the northern edge of the Sargasso Sea. Station S is located at 32° 10′ N, 64° 30′ W, about 21 km southeast of Bermuda within the recirculation region of the North Atlantic anticyclonic subtropical gyre (Worthington 1976, 93). It has been occupied about every two to four weeks over the last 40 years (Jenkins and Goldman 1985, 467). Surface water normally flows into the area from the northeast (Worthington 1976). Horizontal velocities are small, however, according to observations (Michaels and Knap 1996) and modeling studies (Sarmiento and Bryan 1982, Bacastow and Maier-Reimer 1991).

Beginning in 1983, our measurements of dissolved inorganic carbon (DIC), titration alkalinity (TA), and the $^{13}C/^{12}C$ isotopic ratio of DIC, have augmented the standard observations of temperature, salinity, nutrients, and dissolved oxygen at Station S at depths of 1 m and 10 m. These additional measurements, including measurements of nitrate and phosphate, were made on samples of water obtained from approximately monthly hydrographic casts and returned to our laboratory for analysis. They pertain to the surface mixed layer, which almost always extended below the collection depth of the samples employed in this study. The full data set used in this study (years 1983 to 1989) is listed numerically in Tables A.1 and A.2. The methods and precisions of the analyses are described by T. J. Lueker et al. (A twelve year record of inorganic carbon variations in surface ocean water near Bermuda, submitted to Global Biogeochemical Cycles, 1997 [hereinafter referred to as Lueker et al., submitted manuscript, 1997]). A partial summary of the data has already been presented by C. D. Keeling (1993), in which Fig. 5 was mislabeled. It should have read mol C m^{-2} day^{-1}.

We have calculated the partial pressure of CO_2 in the mixed layer (pCO$_{2oc}$) from observations of temperature, salinity, DIC, and TA, using the dissociation constants for carbonic acids established by Dickson and Millero (1987) and the iterative procedure of Bacastow (1981). We reduced all pCO$_{2oc}$ data by 10 ppm in order to agree closely with direct measurements of pCO$_{2oc}$ with gas equilibrators during an intercalibration cruise in the Sargasso Sea in 1987 (Lueker et al. submitted manuscript 1997). This correction is within the uncertainties associated with the use of a particular set of dissociation constants for the calculation of pCO$_{2oc}$ (Millero et al. 1993).

From data supplied by A. Knap (personal communication) we have estimated the location of the base of the surface mixed layer at the time of each hydrographic cast as the depth, which differs by a density difference, $\Delta\sigma_t$, of 0.125 kg m^{-3} (Levitus 1982) from the surface density. We verified that our estimates are reasonable by comparing them visually with plots of density versus depth.

The average seasonal cycles of the principle quantities of our study from 1983 through 1989 are presented in Figures 1 to 5. Plotted are near-surface water temperature, T_{oc}, mixed-layer depth, MLD, DIC normalized to the observed average salinity of 36.452 (on practical salinity scale), sDIC, the stable isotopic ratio of DIC, $^{13}\delta_{oc}$, and near-surface CO_2 partial pressure, pCO$_{2oc}$. All of the cycles shown were obtained by compositing the time series of each parameter into a one year sequence without regard for interannual variability.

Harmonic functions, also plotted, were derived from the same data by least square fits, as described below. Also shown in Figure 5 is a harmonic function for the CO_2 partial pressure in the atmosphere, pCO_{2atm}, estimated for the Bermuda area from measurements at La Jolla, California (Keeling et al. 1989). These data exhibit approximately the same seasonal cycle as air near Bermuda (WMO/WDCGG 1992).

Sea-surface temperature (Figure 1) varies from 19°C in March to 28°C in August. This 9°C range is one of the largest found anywhere in the world oceans (Pickard and Emery 1982, 40). In contrast, salinity (not shown, for data see Table A.1) reveals only a small irregular seasonal cycle with slightly higher values in winter than in summer. The thickness of the mixed layer shows a large seasonal variation (Figure 2). A very shallow layer is normally found from May through September, during the time of high isolation. After September the layer gradually deepens as a result of convection and of intensified mixing induced by stronger winds. A maximum depth, depicted by the harmonic function as being about 160 m, is found during February, near the time when the surface water is coldest. As the time of maximum depth approaches, the thickness of the mixed layer becomes highly variable because the intermittent convective processes act on a weak density gradient near the base of the layer. Beginning in early spring the layer shoals, but the transition from a deep water layer to a shallow one is highly erratic.

Salinity-normalized inorganic carbon, sDIC, (Figure 3) and the reduced isotopic ratio, $^{13}\delta_{oc}$ (Figure 4), show distinct seasonal cycles about averages of 2030 μmol kg^{-1} and 1.53‰, respectively. To aid in the comparison of these two parameters, the plotted isotopic data are inverted, so that higher values on plots of both sDIC and $^{13}\delta_{oc}$ reflect the addition to the DIC pool of organic carbon with a low isotopic ratio. Maxima in sDIC and in negative $^{13}\delta_{oc}$ both occur in spring, nearly coincident with lowest temperatures and greatest mixed-layer depth. As the mixed layer subsequently shoals, sDIC and negative $^{13}\delta_{oc}$ decrease until both reach minima around October. In contrast, titration alkalinity, normalized to a salinity of 36.452, (not plotted) shows no appreciable seasonality around an average of 2380 μEq kg^{-1}, during the analysis period.

The seasonal cycle of pCO_{2oc} (Figure 5) depends strongly on variations in temperature and DIC, but only weakly on salinity and titration alkalinity because of small seasonal variability of the latter two. From December to April high DIC and correlated low temperatures have opposing influences on pCO_{2oc}, causing it to be nearly constant (see also C. D. Keeling 1993). The winter average of pCO_{2oc} is approximately 50 ppm below pCO_{2atm}, also plotted in Figure 5, indicating that the waters near Bermuda are a sink for atmospheric CO_2 during this season. In May a sharp increase in temperature, only partially offset by decreasing DIC, causes pCO_{2oc} to rise, exceeding atmospheric levels by early summer. A maximum approximately 40 ppm above pCO_{2atm} occurs in July, earlier than the temperature maximum because the effect of increasing temperature later in the season is overridden by decreasing DIC. Beginning in September, cooling reinforces the decrease of pCO_{2oc} caused by decreasing DIC, until the flat winter minimum again is reached.

Concentrations of the inorganic nutrients, nitrate and phosphate, are low throughout the year, with highest values in winter or early spring (Michaels et al. 1994a). Nitrate concentrations average about 0.05 μmol kg^{-1} and phosphate about 0.04 μmol kg^{-1}, only slightly above the detection level. Submixed layer data are presented by Lueker et al. (Submitted manuscript 1997).

5. HARMONIC FITTING

A harmonic function,

$$H = \sum_{k=1}^{m} [a_k \sin(2\pi kt) + b_k \cos(2\pi kt)] + H_o,$$ (7)

was fitted by the method of least squares through the composited data of temperature, T_{oc}, sDIC, $^{13}\delta_{oc}$, pCO_{2oc}, and mixed-layer depth, MLD, as a function of time of the year. In equation (7), a_k, b_k, and H_o represent constants that differ for each parameter while t denotes the time expressed in years. Harmonics with periods of 12 months and 6 months (m = 2) were used, except for MLD for which a 4 month harmonic (m = 3) was also included in the fit. The obtained parameters of all fits, including the coefficient of variation, R^2, are shown in Table 1. The constant term, H_o, defines the annual average for the specified parameter.

Isotopic data show considerable scatter, much of which reflects occasions when the data at 1 m and 10 m are not in good agreement. To reduce the influence of this scatter, data that differ between these depths by more than 0.05‰ on the same date were excluded from the fit, even though these data may reflect real differences, for example caused by short-term inhomogeneity in the mixed layer. This criterion of exclusion, which is approximately the minimum error expected from sampling and analysis in measuring the isotopic ratio of atmospheric CO_2 (Mook et al. 1983), led to the removal of about 15% of the data prior to fitting.

6. DESCRIPTION OF THE SEASONAL MODEL

6.1 OUTLINE OF THE MODEL

We have chosen a three-box model, presented schematically in Figure 6, to represent the seasonal cycle of carbon in the ocean near Bermuda. The middle box represents the mixed layer (ml) which lies in direct contact with the air-sea interface. It is overlain by an atmospheric box (atm) and directly underlain by a box representing the water in the "submixed layer" (lb, for "layer below"). The upper- and lower-most boxes are of indeterminate size. They are involved in the model only to establish boundary conditions for the mixed layer. The boundary between the two oceanic boxes moves up and down as specified by the harmonic function that prescribes the mixed-layer depth. Each oceanic box contains, in addition to the chemical pool of DIC, the carbon in organic matter. The concentration of DIC normalized to constant salinity, sDIC, is assumed to be homogeneous within the mixed layer and to be linearly increasing with depth immediately below in the submixed layer. The organic carbon pools of the mixed and submixed layers do not distinguish between dissolved and particulate phases (DOC and POC, respectively); they are considered only with respect to their influence on DIC.

In the model, we have identified four processes in the mixed layer that affect sDIC:

(1)   Air-sea exchange of CO_2 that transfers carbon between the atmosphere and the ocean;
(2)   transport of carbon from below by vertical turbulent diffusion;
(3)   vertical entrainment of water and its content of DIC from below into the mixed layer whenever this layer deepens;
(4)   net exchange of carbon within the mixed layer between the pools of DIC and organic matter caused by photosynthesis, respiration, grazing, and remineralization. When we refer to model results, this net exchange will be called "net biological exchange flux" or simply the "biological flux." It is equivalent to net community production, as defined in Section 2, above.

We have not taken into account possible formation of $CaCO_3$ by organisms. There was no indication that this process is quantitatively important to average seasonal changes in DIC near Bermuda based on the small seasonal variations seen in the alkalinity data (Bates, Michaels, and Knap 1996).

We also have not included air injection by bubble formation of breaking waves in the model, because buffering of the carbon system in seawater causes this process to be unimportant for CO_2, although it should be considered in determining the concentrations of oxygen and the noble gases dissolved in seawater (see Spitzer and Jenkins 1989, R. F. Keeling 1993).

We assume that part of the organic matter produced in the mixed layer is transferred to the layer below. Most of this material is subsequently remineralized to DIC and partly transferred back into the mixed layer by vertical diffusion and entrainment. Organic carbon transfers are not modeled explicitly, but are assumed to close the carbon cycle of the combined oceanic boxes.

We do not take into account transport associated with lateral advection or upwelling or downwelling because we lack observations to assess their importance to DIC in the mixed layer. We will, however, compare our model results with carbon fluxes predicted by the three-dimensional ocean tracer transport model of Bacastow and Maier-Reimer (1991), which includes these processes.

The observed changes in salinity-normalized DIC, denoted by $\Delta sDIC_{oc}(t)$, and in isotopic ratio of DIC, denoted by $\Delta^{13}\delta_{oc}(t)$, in the mixed layer for a time-interval from t_{-1} to t are assumed to consist of the sum of changes produced by the four processes identified above, consistent with equations (3) and (4):

$$\Delta sDIC_{oc}(t) = sDIC_{oc}(t) - sDIC_{oc}(t_{-1}) \tag{8}$$
$$= \Delta sDIC_{ex}(t) + \Delta sDIC_{diff}(t) + \Delta sDIC_{ent}(t) + \Delta sDIC_{bio}(t)$$

and

$$\Delta^{13}\delta_{oc}(t) = \Delta^{13}\delta_{oc}(t) - \Delta^{13}\delta_{oc}(t_{-1})$$
$$= \Delta^{13}\delta_{ex}(t) + \Delta^{13}\delta_{diff}(t) + \Delta^{13}\delta_{ent}(t) + \Delta^{13}\delta_{bio}(t) \tag{9}$$

where, in both equations *ex* refers to gas exchange, *diff* to vertical diffusion, *ent* to entrainment, and *bio* to exchange of carbon between organic matter and DIC. All terms, as indicated, are explicitly functions of time.

As noted in Sections 1 and 3, we employ two strategies to calculate the contribution of the biological processes, $\Delta sDIC_{bio}$, which constitutes the only concentration term that cannot be estimated from empirical relationships.

In the first method each of the first three process terms listed in equations (8) and (9) ($\Delta sDIC_{ex}$, $\Delta sDIC_{diff}$, $\Delta sDIC_{ent}$, and the corresponding terms for $\Delta^{13}\delta$) are evaluated individually for each time step by means of empirical relationships explained in the following subsections. The isotopic change caused by biological processes, $\Delta^{13}\delta_{bio}$, is then solved for as a difference from the other terms in equation (9), with the sum, $\Delta^{13}\delta_{oc}(t)$, evaluated from the harmonic representation of the observations expressed by equation (7). The difference, $\Delta^{13}\delta_{bio}$, is then used to calculate $\Delta sDIC_{bio}$, where account is taken of isotopic fractionation of carbon occurring during photosynthesis. In this case all four contributions on the DIC balance in the mixed layer can be computed and their temporally integrated sum, a quantity that we will call $sDIC_{calc}$, can be compared to the observed seasonal variations of $sDIC_{oc}$ to check the consistency of our computations.

In the second method a quantity, which we will call $\Delta sDIC'_{bio}$, to distinguish it from the corresponding quantity determined by the first method, is determined from the observed change, $\Delta sDIC_{oc}(t)$, by subtracting the calculated changes ascribed to the other three fluxes of equation (8), i.e.

$$\Delta sDIC'_{bio}(t) = \Delta sDIC_{oc}(t) - \Delta sDIC_{ex}(t) - \Delta sDIC_{diff}(t) - \Delta sDIC_{ent}(t). \tag{10}$$

This second approach is identical to previously employed techniques for estimating net community production from inorganic carbon data (Oudot 1989, Karl, Tilbrook, and Tien 1991, Chipman, Marra, and Takahashi 1993, Robertson and Watson 1993). Our principal interest is in the first method, which takes isotopic variations into account, but the second method provides a useful check on the first method, as mentioned in Section 1. The model calculations to depict the seasonal cycle of carbon near Bermuda were carried out during a single annual cycle, using a constant time step, Δt, of one day.

In the following subsections, we present details of the computational schemes used for these model calculations. A more detailed description of the formulas is furnished in Appendix A. The input data and parameters used for the model calculation are summarized in Table 2.

6.2 AIR-SEA GAS TRANSFER

The net exchange flux of CO_2 gas across the air-sea interface, F_{ex}, at each time step, is evaluated by the expression

$$F_{ex}(t) = k_{ex}(t_{-1})\ (pCO_{2atm}(t_{-1}) - pCO_{2oc}(t_{-1})), \tag{11}$$

where pCO_{2atm} denotes the partial pressure of CO_2 in the atmosphere immediately above the sea surface, and pCO_{2oc} the partial pressure at the sea surface in the ocean. The only observations of pCO_{2atm}, including its isotopic ratio and pertaining to nearly the same latitude as Bermuda, which are available to establish the seasonal cycle during the period of investigation there, were obtained at La Jolla, California, at 32.9°N on the Pacific coast of North America. Because the seasonal cycle in pCO_{2atm}, according to a model calculation (Heimann, Keeling, and Tucker 1989, 292), varies only slightly with longitude in midlatitudes, we deem the data for La Jolla to be adequate to represent pCO_{2atm} at Station S. This assumption has been verified for the years from 1989 to 1991 with data obtained at island station Bermuda East (WMO/WDCGG 1992). A two-harmonic fit ($m = 2$ in equation (7)), adjusted for secular change to a datum of 1987, was used to represent these atmospheric observations in the model. We disregard the small difference between partial pressure and fugacity, the latter being regarded from a thermodynamic point of view as the exact parameter governing air-sea exchange (Weiss 1974).

The gas-exchange coefficient, k_{ex}, is expressed as a product of the CO_2 gas solubility, α_s, and the piston velocity, P_v (Broecker and Peng 1982, 122). Solubility is computed as a function of salinity and temperature using the formulation of Weiss (1974). Salinity is assumed to be constant at a value, S_o equal to 36.452, the average salinity of our time series. We adopted the formula of Heimann and Monfray (1989) (based on Liss and Merlivat [1986]) for the wind-speed dependency of the piston velocity, adjusted to agree with radiocarbon data by increasing k_{ex} by a constant factor, γ, of 1.7447. Climatological monthly average wind-speed data were taken from Musgrave, Chou, and Jenkins (1988), who compiled them from Isemer and Hasse (1985) for a 1° square centered at 32.5°N, 64.5°W. The wind data are plotted in Figure 7. The formulas used to compute k_{ex} are given in Appendix A.1.

The corresponding flux of $^{13}CO_2$, $*F_{ex}$ is calculated by an expression similar to equation (11) but allowing for isotopic fractionation (see Heimann and Keeling [1989, 264, equation (5.41)]). Thus,

$$*F_{ex}(t) = k_{ex}(t_{-1})(R_s/r_s)\alpha_{am}[pCO_{2atm}(t_{-1})r_{atm}(t_{-1}) -$$
$$pCO_{2oc}(t_{-1})\alpha_{eq}(t_{-1})r_{oc}(t_{-1})], \tag{12}$$

where R_s denotes the $^{13}C/(^{12}C + ^{13}C)$ ratio of the isotopic standard PDB, r_s the corresponding $^{13}C/^{12}C$ ratio, r_{atm} the $^{13}C/^{12}C$ ratio of CO_2 in the atmosphere, and r_{oc} the $^{13}C/^{12}C$ ratio of DIC in the mixed layer. The factor α_{am} denotes the kinetic $^{13}C/^{12}C$ fractionation for CO_2 uptake by the surface ocean water, and α_{eq} the equilibrium isotopic fractionation factor of gaseous CO_2 with respect to DIC (see Appendix A.1).

The seasonal cycle of r_{atm} is evaluated from observations of $^{13}C/^{12}C$ in atmospheric CO_2 at La Jolla, California, from 1978 to 1989, adjusted to a datum of 1987, as was pCO_{2atm}. A two-harmonic fit of the reduced ratio, $^{13}\delta_{atm}$, was utilized directly in equation (A.6) of Appendix A.1.

Over each time step, both the ^{13}C flux, $*F_{ex}(t)$, and the $^{12}C + ^{13}C$ flux, $F_{ex}(t)$, are assumed to be constant. For the calculation of changes in concentration, $\Delta sDIC_{ex}$ and in isotopic ratio, $\Delta^{13}\delta_{ex}$, and the fluxes are assumed to be added to the mixed layer, defined by its depth at the beginning of the time step (see equations (A.5) and (A.9) of Appendix A.1).

6.3 VERTICAL TURBULENT DIFFUSIVE TRANSPORT

The change in sDIC in the mixed layer caused by diffusive transport at its base is calculated using an eddy diffusion formulation. The vertical flux of DIC into the mixed layer by turbulent diffusion, F_{diff}, is thus defined by the expression:

$$F_{diff}(t) = K_z(t_{-1})(dsDIC/dz)_{lb}\,\rho_o, \tag{13}$$

where K_z denotes the vertical diffusion coefficient at the base of the mixed layer, $(dsDIC/dz)_{lb}$ denotes the vertical gradient in $sDIC$ immediately below the mixed layer, assumed to be constant with depth and time, and ρ_o denotes the seawater density. The density, ρ_o, appears in equation (13) only to give the correct dimension to F_{diff} and hence only a nominal constant value was used (see Table 2). The concentration of sDIC is assumed to be continuous at the boundary between the mixed layer and submixed layer.

Coefficient K_z is determined according to a formula of Denman and Gargett (1983):

$$K_z = 0.25\varepsilon(N)^{-2}, \tag{14}$$

where ε denotes the turbulent kinetic energy dissipation, assumed to be constant immediately below the mixed layer, and N the buoyancy (Brunt-Väisälä) frequency, an expression of the water-column stability that depends on the vertical gradient of the time varying seawater density, $\rho(t)$ (see Appendix A.2, equation A.10). The annual cycle of K_z, shown in Figure 8, was evaluated from separate density profiles, based on temperature and salinity data for each occupation of Station S from 1983 to 1989, whether or not sampled as part of our study. As noted above, a harmonic function with three terms ($m = 3$ in equation (7)) was assumed. The coefficients of the fit are given in Table 1. We did not attempt a fit using higher harmonics, and thus probably have underestimated K_z in late winter and early spring.

With the formation of a strong pycnocline in May, K_z decreases rapidly to low and almost negligibly varying values around $0.2 \cdot 10^{-4}$ m^2 s^{-1} that prevail throughout the summer. These values agree well with estimates based on microstructure measurements (($0.3 \cdot 10^{-4}$ m^2 s^{-1}) or less according to Denman and Gargett [1983]) or from an open ocean tracer release experiment ($0.1 \cdot 10^{-4}$ m^2 s^{-1} according to Ledwell, Watson, and Law [1993]), both obtained in the main thermocline. With the erosion of the pycnocline in autumn, the diffusion coefficient gradually increases until it becomes large and erratic in February. For late winter, calculated K_z may attain values 10 times those normally found for midsummer (around $2 \cdot 10^{-4}$ m^2 s^{-1}). These values are perhaps too high, but lie within the range of estimates based on bulk mixed-layer models ($1 \cdot 10^{-4}$ m^2 s^{-1} according to Musgrave, Chou, and Jenkins [1988]) and observations in the Northeast Pacific during the passage of a storm ($10 \cdot 10^{-4}$ m^2 s^{-1} according to Large, McWilliams, and Niiler [1986]).

The vertical turbulent diffusive flux of ^{13}C, $*F_{diff}$, is computed in a manner identical to F_{diff}. (See Appendix A.2.)

Because we lack the observations to establish the variability in sDIC below the mixed layer, we assumed a constant vertical gradient of 0.45 ± 0.15 μmol kg^{-1} m^{-1} (z axis positive downward), derived from four vertical profiles of sDIC that were measured in the course of the six years of observations (see Lueker et al. submitted manuscript 1997). Isotopic measurements derived from the same four vertical profiles are also too few to estimate the time variability of $^{13}\delta$ below the mixed layer. They provide a rough estimate for the average gradient of -0.003 ± 0.001‰ m^{-1} ($^{13}\delta$ becoming more negative downward).

Because this latter estimate is highly uncertain, we calculated the isotopic gradient by means of the three-dimensional ocean tracer transport model of Bacastow and Maier-Reimer (1991) (see Appendix B). As shown in Table 3, the vertical gradients of DIC and $^{13}\delta$, according to this model, are caused by remineralization of organic carbon, the buildup of DIC with a low isotopic signature in the upper ocean from fossil fuel combustion (Suess effect) and by the temperature-dependency of the CO_2 solubility and of the isotopic fractionation of CO_2 during air-sea exchange (see also Section 2). We adjusted the computed overall gradient of DIC slightly in order to match the observed value of 0.45 μmol kg^{-1} m^{-1} (see Appendix B for details). The correspondingly adjusted isotopic gradient was found to be -0.0021‰ m^{-1}, somewhat smaller than the observed gradient of -0.003 ± 0.001‰ m^{-1}, but within the range of uncertainty in the observations.

Changes in sDIC and in $^{13}\delta_{oc}$ from vertical turbulent diffusion are computed by an expression analogous to that used in determining the changes from air-sea exchange (see equations (A.11) and (A.14) of Appendix A.2).

6.4 VERTICAL ENTRAINMENT

The passive transport of DIC by entrainment of water into the mixed layer during times of deepening, F_{ent}, is computed assuming that the additional water mass, defined by the increased mixed-layer depth from beginning to end of an episode of entrainment, is at that time instantly incorporated into the mixed layer. We regard the recurrence interval of entrainment, Δt_{ent}, to be independent of the time step of the model, Δt (see Appendix A.3).

We estimated Δt_{ent} to be 8 days on the basis of limited evidence for actual entrainment events from a study at Ocean Weather Station P in the North Pacific Ocean (Large, McWilliams, and Niiler 1986, Figure 13, 1535). We smoothed the predicted change in composition by assuming that a fraction $\Delta t / \Delta t_{ent}$ of each episode occurs during one time step of the model. In our simulation, entrainment occurs from day 191 to day 51, inclusive, of the next year (cf. Figure 2).

We cannot compute entrainment more realistically because we lack a sufficiently detailed knowledge of short-term vertical variations in properties near the base of the mixed layer. Thus, we did not attempt to estimate entrainment arising from short-term deepening of the mixed layer during that part of the annual cycle when shoaling predominates. Because we provide in the model for vertical transport by eddy diffusion (see the previous subsection), we have accounted to some degree for vertical transport in DIC arising from short-term variations in the same properties that govern entrainment. Details of the calculations of the changes in concentration and isotopic ratio resulting from entrainment are given by equations (A.27) and (A.32) respectively in Appendix A.3.

6.5 BIOLOGICAL EXCHANGE: NET COMMUNITY PRODUCTION

Fractionation during photosynthesis is assumed to be analogous to a Raleigh process (Mook, personal communication), with a single sink and no source:

$$\frac{r_{end}}{r_{init}} = \left[\frac{sDIC_{end}}{sDIC_{init}} \right]^{\alpha_{org} - 1}, \tag{15}$$

where r_{init} denotes the $^{13}C/^{12}C$ ratio in the DIC pool at the beginning of the process and r_{end} at the end, and where $sDIC_{init}$ and $sDIC_{end}$ denote the corresponding sDIC concentrations. The fractionation factor of photosynthesis, α_{org}, in the mixed layer is defined in terms of reduced isotopic ratios by:

$$\alpha_{org} = {}^{13}\delta_{org} - {}^{13}\delta_{oc} + 1, \tag{16}$$

where $^{13}\delta_{org}$ denotes the average $^{13}\delta$ value of phytoplankton and $^{13}\delta_{oc}$ that of the DIC pool in the mixed layer.

We evaluate $^{13}\delta_{org}$ according to the formula of Rau, Takahashi, and Des Marais (1989), who related the observed variability in $^{13}\delta$ of phytoplankton to differences in CO_2 solubility, based mainly on data from the Atlantic and Southern Oceans. From these data they found empirically that:

$$^{13}\delta_{org} = (-0.8 \, [CO_2(aq)] - 12.6) \times 10^{-3}, \tag{17}$$

where $[CO_2(aq)]$ denotes the local concentration of dissolved CO_2 in seawater in μmol kg^{-1}. We then calculate α_{org} by equation (16), where $^{13}\delta_{oc}$ is given by the harmonic fit to our observations. Because the variability of $[CO_2(aq)]$ is small over the temperature range observed at Station S, $^{13}\delta_{org}$ is found to vary over only a narrow range ($\pm 0.2\%o$) from its average of -20.499‰, while α_{org} also varies over a narrow range from its average departure from unity of -22.026‰ (where the difference, 1.527‰, is the average of $^{13}\delta_{oc}$, as given in

Table 1). This computed value of $^{13}\delta_{org}$ agrees well with observations by Druffel et al. (1992) obtained at Station S. They found the average $^{13}\delta$ of particulate organic carbon (POC) in the top 100 meters to be -20.8 ± 0.5‰.

The change in sDIC in the mixed layer resulting from biological processes, $\Delta sDIC_{bio}$, during time step, Δt, (from t_{-1} to t) is the difference between $sDIC_{init}$ and $sDIC_{end}$, and is computed by solving equation (15) for $sDIC_{init}$ and then subtracting $sDIC_{init}$, hence:

$$\Delta sDIC_{bio}(t) = sDIC_{init} \left[\frac{^{13}\delta_{end}(t) + 1}{^{13}\delta_{init}(t) + 1} \right]^{\left(\frac{1}{\alpha_{org}} - 1 \right)} - sDIC_{init}, \qquad (18)$$

where we substitute $(^{13}\delta + 1)r_s$ for r according to equation (1) to obtain expressions in $^{13}\delta$ notation, and where:

$$^{13}\delta_{end}(t) = {}^{13}\delta_{oc}(t). \qquad (19)$$

The terms $sDIC_{init}(t)$ and $^{13}\delta_{init}(t)$ refer to $sDIC(t_{-1})$ and $^{13}\delta(t_{-1})$, after adjustments that take account of air-sea exchange, vertical diffusion, and entrainment, i.e.:

$$sDIC_{init}(t) = sDIC(t_{-1}) + \Delta sDIC_{ex}(t) + \Delta sDIC_{diff}(t) + \Delta sDIC_{ent}(t) \qquad (20)$$

and

$$^{13}\delta_{init}(t) = {}^{13}\delta_{oc}(t_{-1}) + \Delta^{13}\delta_{ex}(t) + \Delta^{13}\delta_{diff}(t) + \Delta^{13}\delta_{ent}(t), \qquad (21)$$

consistent with equations (8) and (9) respectively.

6.6 CALCULATION OF VERTICALLY INTEGRATED RATES OF CHANGE

As an alternative to expressing the annual carbon cycle in surface waters in terms of changes in concentration, we consider the rate of change in sDIC for each component of the carbon cycle, F_i, vertically integrated over the mixed layer. The seasonal cycles of the F_i differ in pattern from the former rates because of the variable depth of the mixed layer.

Per unit area, and in the approximation of small time steps, these vertically integrated fluxes are given in our model by the expression:

$$F_i(t) = \frac{\Delta sDIC_i(t) \cdot MLD(t_{-1})}{\Delta t}, \qquad (22)$$

where $\Delta sDIC_i(t)$ denotes the change in sDIC caused by process i during the time step, $MLD(t_{-1})$ denotes the mixed-layer depth at the beginning of the time step, and Δt denotes the duration of the time step.

As a check on the consistency of all four calculated process fluxes, we compare their sum:

$$F_{calc}(t) = \Sigma F_i(t), \qquad (23)$$

21

to the vertically integrated rate of change in sDIC based on direct observations:

$$F_{obs}(t) = \frac{\Delta sDIC_{oc}(t) \cdot MLD(t_{-1})}{\Delta t}. \tag{24}$$

This flux is to be distinguished from the total rate of change in sDIC, F_{tot}, which depends on changes in both water depth and concentration within the mixed layer. In the notation of differential calculus

$$F_{tot} = d[MLD(t) \cdot sDIC_{oc}(t)]/dt \tag{25}$$

whereas F_{obs} represents only that part caused by changes in concentration, disregarding the change in volume in the water column. Thus, again in differential notation

$$F_{obs}(t) = MLD(t) \, (dsDIC_{oc}/dt) \tag{26}$$

$$= F_{tot} - sDIC_{oc}(t) \, (dMLD(t)/dt), \tag{27}$$

where the second term on the right hand side of equation (26) expresses the mass flux through the base of the mixed layer resulting from shoaling and deepening of this layer.

Because the mixed-layer depth at Station S varies over a large range, between 15 m and 160 m in our harmonic representation based on observations, the total flux F_{tot} is much larger than $F_{obs}(t)$. We are primarily interested, however, in the integrated rates of change caused by local changes in concentration, and thus we focus on the quantity $F_{obs}(t)$. The second term on the right hand side of equation (27) is important, however, to close the mass balance of the system.

7. RESULTS OF THE SEASONAL MODEL

7.1 SEASONAL VARIATIONS OF DISSOLVED INORGANIC CARBON

The rates of change in the salinity-normalized DIC concentration calculated by the model over the course of the annual cycle, are shown in Figure 9 individually for gas exchange, vertical turbulent diffusive transport, vertical entrainment, and net biological exchange.

Substantial variations in rates are exhibited by gas exchange and biological processes during most of the year. Somewhat smaller variations are shown by vertical diffusion, whereas entrainment causes only minor variations.

The variations in gas-exchange rate mainly reflect the CO_2 pressure gradient at the air-sea boundary. During the cooler part of the year, when the CO_2 partial pressure in the water (pCO_{2oc}) is below the atmospheric partial pressure (pCO_{2atm}), gas exchange causes sDIC to increase at a nearly steady rate of about 0.08 μmol kg^{-1} day^{-1} from November through most of May. Subsequently, during the summer months, when the seawater temperature is high and pCO_{2oc} exceeds pCO_{2atm}, the net flux at the air-sea interface is reversed, and sDIC decreases, attaining a maximum rate of -0.11 μmol kg^{-1} day^{-1} in mid-August.

According to our model calculations, biological processes remove carbon from the inorganic carbon pool during most of the year, reflecting positive net community production. Only from mid-November until the beginning of March is there a net return of organic carbon to the inorganic pool. In spring when primary production is maximal (Menzel and Ryther 1960, Michaels et al. 1994a) the rate of inorganic carbon removal, $\Delta sDIC_{bio}/\Delta t$, attains 0.24 μmol kg^{-1} day^{-1}, (3.0 mg C m^{-3} day^{-1}). During the period of high production ($\Delta sDIC_{bio}/\Delta t$ 0.20 μmol kg^{-1} day^{-1}) from early April to mid-June, the mixed layer, as prescribed by the harmonic fit, shoals from 120 m to 20 m. For this period (66 days), the average removal rate, integrated over the mixed layer, is 0.18 gC m^{-2} day^{-1}. After a period of low inorganic carbon removal during summer, a secondary maximum in removal rate of 0.20 μmol kg^{-1} day^{-1} is observed in late September. The existence of this secondary maximum rate is consistent with the autumnal plankton bloom found by Menzel and Ryther (1960) from measurements of primary production, although the bloom occurred about one month later. Similarly, Lohrenz et al. (1992) observed an increase in primary production in July 1989 and from October to December 1990.

The onset of negative net community production (return of inorganic carbon to the mixed-layer pool) in mid-November occurs at a time when primary production is decreasing and when the mixed layer is deepening rapidly, thereby entraining waters with an excess of respiration over net primary production. Net community production in the mixed layer remains negative until the beginning of March, even though primary production increases considerably already in January (Michaels et al. 1994a). During this period the mixed layer includes a large amount of water lying below the euphotic layer. The vertically integrated respiration in these waters is obviously large enough to compensate for the primary production until the beginning of March, when shoaling of the mixed layer combined with increased light availability leads to positive net community production in the mixed layer. These observations are consistent with Sverdrup's critical depth hypothesis (Sverdrup 1953), which relates the onset of the spring bloom with an increase in the water-column stability.

Variations in sDIC concentration from vertical diffusive transport, $\Delta sDIC_{diff}/\Delta t$, are found by the model to be small except in spring, a maximum of 0.09 μmol kg^{-1} day^{-1} being observed in mid-May. The contribution from vertical entrainment, compared to the other contributions, is small, even during the period of maximum deepening of the mixed layer in December and January when $\Delta sDIC_{diff}/\Delta t$ attains 0.03 μmol kg^{-1} day^{-1}.

The rate of change in DIC by biological processes, $\Delta sDIC_{bio}/\Delta t$, as plotted in Figure 9, was computed from isotopic data by equation (18), above. Alternatively, as noted above in Subsection 6.1, this rate of change can be determined from the observed change in sDIC in the mixed layer, $\Delta sDIC_{obs}$, merely by subtracting from it, for each time step, the calculated changes ascribed to the other three fluxes as given by equation (10). The outcomes of the isotopic and alternative ways of calculating, distinguished respectively by the symbols, $\Delta sDIC_{bio}$ and $\Delta sDIC'_{bio}$, are shown in Figure 10. Differences between the two methods, which reach 0.10 μmol kg^{-1} day^{-1} in autumn, reflect a bias toward lower values during shoaling of the mixed layer when the rate of change is calculated by differences in sDIC, and higher values during deepening, suggestive of errors contributed by the calculated fluxes by diffusion or entrainment. Nevertheless, the two ways of calculating $\Delta sDIC_{bio}/\Delta t$ show a similar overall pattern, a rapid decline from a maximum in early winter to low negative values in spring, and then small variations until the rate increases again in autumn.

7.2 CALCULATED SEASONAL CYCLE OF DISSOLVED INORGANIC CARBON

The seasonal cycle of sDIC, calculated as the cumulative sum of increments for each process via equation (8), and designated $sDIC_{calc}$ (see Subsection 6.1), is compared in Figure 11 to the observed seasonal cycle of $sDIC_{oc}$ of equation (8). The starting value of $sDIC_{calc}$ is set so that the annual average agrees with $sDIC_{oc}$. The curves differ in pattern as a direct result of the integrated differences between $\Delta sDIC_{bio}$ and $\Delta sDIC'_{bio}$, shown in Figure 9. The curve of $sDIC_{calc}$ has nearly the same amplitude as that of $sDIC_{obs}$, but lags in phase and fails to close exactly in the annual cycle, although it should. The final value for December 31 is about 1.8 μmol kg^{-1} lower than the initial value for January 1. This lack of closure is presumably a result of uncertainties in the calculations of the different terms in equations (8) and (9).

7.3 SEASONAL VARIATIONS OF VERTICALLY INTEGRATED RATES OF CHANGE

Seasonal variations in the rates of change of sDIC for the different processes integrated over the mixed layer, F_i (defined by equation (22)), are presented in Figure 12. The relative contributions of the integrated rates at any given time are the same as the changes in sDIC shown in Figure 9, but from season to season the integrated rates are different in absolute terms, because of variations in mixed-layer depth. In particular, all integrated changes, F_i, are relatively small during summer when the mixed layer is shallow.

During March and April the net biological exchange flux, F_{bio}, shows a pronounced maximum in removal of DIC from the mixed layer (-0.27 gC m^{-2} day^{-1}). A secondary maximum removal rate occurs in mid-October (-0.07 gC m^{-2} day^{-1}). Because the mixed layer is deeper in the spring than in autumn, the spring maximum occurs earlier and is larger. Also, the autumnal maximum is later and smaller than the respective maximum for $\Delta sDIC_{bio}/\Delta t$, and thus more nearly in accord with the findings of Menzel and Ryther (1960) and Lohrenz et al. (1992), as noted in Subsection 7.1. The rates of increase in sDIC in the mixed layer caused by gas exchange and diffusion also reach extremes in spring (0.14 and 0.11 gC m^{-2} day^{-1}, respectively). Vertical diffusive transport, F_{diff}, is close to zero in late summer and during most of autumn. Gas exchange, F_{ex}, is also small in summer but increases again in late autumn. The vertical entrainment flux, F_{ent}, is small or zero at all times.

Seasonal variations of the vertically integrated rate of change of sDIC, summed from all components and denoted by F_{calc}, are compared with the rate based on direct observations, F_{obs}, in Figure 13. Disagreements between the curves are seen to be consistent with those for $\Delta sDIC/\Delta t$, shown previously in Figure 10.

7.4 TEMPORALLY INTEGRATED CHANGES IN CONCENTRATION

The net changes in concentration of sDIC in the mixed layer for each process are presented in Table 4 for the annual cycle and summed over the shoaling period, from mid-February until early July (days 51 through 190 of the annual cycle, see Figure 2), and for the deepening period thereafter. During the year, biological exchange removes 0.42 gC m^{-3} from the mixed layer. Most of this change, 0.29 gC m^{-3}, occurs during the shoaling period. Gas exchange and diffusive transport cause DIC to increase in the mixed layer for both periods. The increases are larger during shoaling. For entrainment, a small increase occurs during deepening. Over the annual cycle the model calculation, $sDIC_{calc}$, indicates a net addition from all processes of -0.02 gC m^{-3}, whereas the net change inferred directly from the observed seasonal cycle of $sDIC_{oc}$, of course, is identically zero. During shoaling and deepening the calculated changes show somewhat larger discrepancies from the observed changes ($sDIC_{calc}$ minus $sDIC_{oc}$) of 0.09 and -0.12 gC m^{-3}, respectively.

7.5 TEMPORALLY AND VERTICALLY INTEGRATED RATES OF CHANGE

Vertically integrated rates of exchange of the inorganic carbon system in the mixed layer, F_i, are listed in Table 5, arranged in a manner similar to Table 4. Biological exchange removes 19 gC m^{-2} from the mixed-layer DIC pool during shoaling, but returns 8 gC m^{-2} during the deepening period, resulting in a net annual removal of 11 gC m^{-2}. During the annual cycle, air-sea exchange transfers 21 gC m^{-2} yr^{-1} into the mixed layer, nearly equally divided between periods of shoaling and deepening. Vertical diffusive transport over the annual cycle contributes a substantial 15 gC m^{-2}, predominantly during shoaling. Vertical entrainment makes a smaller contribution, as was the case for $\Delta sDIC_{ent}/\Delta t$, but when added to vertical diffusive transport causes the sum to be nearly as large during deepening as during shoaling. Disagreement exists between F_{calc}, the calculated sum of the four component flux terms (see equation (23)), and the observed rate, F_{obs}, (equation (24)), as estimated from changes in concentration. The absolute disagreement is greater during shoaling (7 gC m^{-2}) than during deepening (5 gC m^{-2}).

During the annual cycle the observed vertically integrated rate, F_{obs}, is a positive 27gC m^{-2} yr^{-1}, balanced by an equal apparent loss expressed by the annual integral of the flux associated with the varying mixed-layer depth ($sDIC_{oc}[dMLD/dt]$) (see equation (24) and Table 5). The latter, together with F_{obs} sums to the total rate, F_{tot}, which over the annual cycle is zero by definition. The annually integrated fluxes F_{obs} and $sDIC_{oc}[dMLD/dt]$ are different from zero because the mixed layer shoals at times when the concentration of DIC in the mixed layer, on average, is higher than when it is deepening. The implied imbalance below the mixed layer, if our calculations are correct, must ultimately be compensated for by vertical or horizontal fluxes of inorganic or organic carbon, as discussed below.

8. DISCUSSION

8.1 SUMMARY AND DISCUSSION OF THE SENSITIVITY TESTS

Introduction

The seasonally varying fluxes of carbon predicted for the waters near Bermuda by our model are all influenced to some degree by uncertainties in the relationships used for the calculations. After identifying uncertainties in estimating each flux, we challenge the plausibilities of these relationships by means of sensitivity tests. The results of these tests are summarized in Table 6 with respect to the computed net community production in the mixed layer, expressed in our model by the net biological exchange flux, F_{bio}, and with respect to the computed seasonal variations of salinity-normalized DIC, $sDIC_{calc}$, in Figure 14. Also shown in Table 6 is the lack of annual closure of the computed seasonal cycle in $sDIC_{calc}$, designated as $\Delta sDIC_{calc}$. Results are compared to a standard case, defined as the set of calculations summarized in Tables 4 and 5. Details concerning these tests are presented in Appendix C.

Air-sea exchange

We estimate that the statistical uncertainty for the relationship between measured piston velocity, P_v, and wind speed (equation A.3 in Appendix A) is about 15%, based on the scatter of various estimates (Heimann and Monfray 1989, Figure 6). In addition to random uncertainty, it is likely that the gas-exchange coefficient, computed by the formula of Liss and Merlivat (1986) from monthly average wind speed, is biased on the low side, because this formula, based on wind-wave facility data, prescribes a steeper dependence on wind speed when the instantaneous winds are high (See Appendix A). In an attempt to compensate for this bias, we increased the calculated gas-exchange coefficient by a factor, γ, set equal to 1.7447 (see Subsection 6.2) as proposed by Keeling, Piper, and Heimann (1989b) and similar to the value of 1.7 given by Watson (1993). Lacking better evidence, we estimate that the combined errors in establishing the gas-exchange coefficient, k_{ex}, are of the order of 30%. Thus the net flux of CO_2 across the air-sea interface uptake of atmospheric CO_2, F_{ex}, of the annual cycle is estimated to be 21 ± 6 gC m^{-2} yr^{-1}. We have diagnosed the influence of this uncertainty on the model results as follows.

To assess the effect of the large uncertainty assigned to the piston velocity adjustment factor, γ, this factor was varied between 1.0 and 2.0. We found that any error in computing F_{ex}, which mutually impacts both the concentration and the isotopic ratio of DIC, is almost directly passed on to the net biological exchange flux, F_{bio}. In contrast, the error in specifying the CO_2 pressure gradient at the sea surface produces a lesser error in F_{bio} because of a partial compensation from the isotopic relationship.

Some indication of what should be the correct air-sea exchange flux is afforded by looking at how well the annual cycle in $sDIC_{calc}$ closes in the model. A considerably lower gas-exchange coefficient, consistent with the Liss and Merlivat (1986) specification, produces a less desirable closure than our preferred choice. A confirmation that the gas-exchange rate is varying significantly with season is given by a very poor annual closure in $sDIC_{calc}$ when a constant rate for the year is assumed. On the other hand, neglecting the effect of seasonal variations in temperature on the gas-exchange rate has little impact on the model results, either for F_{bio} or the closure of $sDIC_{calc}$.

If we ignore the disequilibrium between atmospheric and oceanic CO_2 caused by the buildup of industrial CO_2 in the atmosphere (the Suess effect), the isotopic relationships are radically altered with major impacts on the calculation of F_{bio}. A strong indication that we have represented the air-sea gradient in $^{13}\delta$ nearly correctly, is the less desirable annual closure in $sDIC_{calc}$ when the Suess effect is neglected.

We note that the Suess effect produces an isotopic disequilibrium of about 1.1‰, whereas the observational errors in $^{13}\delta$ of atmospheric and oceanic CO_2 are each of the order of 0.05‰. Even if there is some additional uncertainty because of biases between the atmospheric and oceanic isotopic measurements, the isotopic errors involved in the air-sea exchange process are too small to have a significant impact on the calculation of F_{bio}. Therefore, we estimate that the uncertainty of 30% in the estimate for the annual air-sea gas transfer leads to an uncertainty of about the same degree in our estimate of annual net community production.

Vertical transport

Errors in our estimation of vertical transport of DIC across the lower boundary of the mixed layer are not readily established because we cannot be sure that our mathematical formulation of either diffusive transport or entrainment is physically realistic. In addition, our estimations of vertical gradients in both concentration and isotopic ratio of DIC, and of the vertical diffusion coefficient, are uncertain, because they are based on sparse data. Furthermore, our estimation of net community production through the biological flux, F_{bio}, is sensitive to the mixed-layer depth, whose variations for time intervals shorter than about one month are not taken into account. Our estimates of vertical diffusive transport may not be in serious error for the deepening period for which we employ a diffusion coefficient, K_z, of about $(0.2 \cdot 10^{-4} \text{ m}^{-2} \text{ s}^{-1})$ in good agreement with recent observation (Ledwell, Watson, and Law 1993), but over the shoaling period we employ much higher values of K_z, (of the order of $2 \cdot 10^{-4} \text{ m}^{-2} \text{ s}^{-1}$) not supported by direct measurements. We cannot rule out substantial errors in the calculation of transport for this period. Thus, and especially if we allow for systematic biases, relative errors of up to 50% in our transport calculations are possible. Given this estimation of uncertainty, sensitivity tests are especially useful to determine how seriously F_{bio} is affected by errors in this transport.

Doubling the rate of vertical transport by diffusion, or reducing it to zero, leads to proportional changes in F_{bio} equal to about 60% of the changes in vertical transport. The 40% reduction in impact is caused by compensation arising from the isotopic relationships of the model. Our isotopic data, which influence the closure of $sDIC_{calc}$, are evidently inconsistent with these extreme values in vertical transport, because they result in very poor annual closure of $sDIC_{calc}$. Thus, if our isotopic formulation and data are trustworthy, it is unlikely that our computation of transport is in relative error to a degree approaching that of these tests.

If we decrease or increase vertical entrainment by 50% through variation of the entrainment recurrence interval Δt_{ent}, by ± 4 days, respectively (see Appendix A.3), F_{bio} is only slightly affected, because entrainment is only a small part of vertical transport in our model. Furthermore, if we assume a small diffusion coefficient throughout the year

$(0.1 \cdot 10^4 \text{ m}^2 \text{ s}^{-1})$, and at the same time drastically increase entrainment (Δt_{ent} to 60 days) so as to produce approximately the same rate of upward transport of DIC as in our standard calculation, F_{bio} again is only slightly affected. Thus, the mixed-layer model is insensitive to whether the vertical transport is predominantly by eddy diffusion, or by entrainment, as long as the sum is constant.

If the observed vertical gradient of $^{13}\delta$ below the mixed layer is used in the calculations instead of the estimate from the three-dimensional model of Bacastow and Maier-Reimer (1991) F_{bio} is increased by about 30%. This substitution also causes a less desirable annual closure of $sDIC_{calc}$, suggesting that the prediction of the isotopic gradient by the model is closer to correct than the gradient based on our limited observations.

To challenge our harmonic estimation of the mixed-layer depth, we replaced our two-harmonic representation with a hermite spline function (Figure 15), which more closely fits the data. The resulting shallower mixed layer in the spring causes predicted changes in concentration of DIC resulting from air-sea exchange and vertical exchange that appear to be exaggerated, but cause only moderate changes in F_{bio}. Varying the depth of the mixed layer over the entire year by 30 m has only a minor impact on F_{bio}.

In conclusion, we cannot rule out errors as large as 50% in our estimation of vertical transport of DIC, and consequently, errors on the order of 30% in net community production. Nevertheless, when vertical transport in the model is increased or decreased substantially from our standard computations, the model predicts poor closure in the annual cycle of $sDIC_{calc}$. This closure indicates that vertical transport much higher or lower than our estimate is inconsistent with the isotopic relationships of the marine carbon cycle, as taken into account in our model. Also, we did not find the model predictions of biological processes to be particularly sensitive to our representation of the mixed layer.

Biological exchange

The only uncertain parameter in the calculation of the net biological exchange flux, F_{bio}, from $^{13}C/^{12}C$ data (see equation 15) is the fractionation factor for photosynthesis, α_{org}. As described in Appendix C, its uncertainty is estimated to be about $\pm 3\%_o$. Variations in the isotopic fractionation factor by this amount showed only little impact on the results of the model.

Taking into account all of the sensitivity tests described above, we estimate that the uncertainty in F_{bio} is of the order of 50%. We base this estimate on the uncertainty of the air-sea exchange and of the vertical processes and their influence on F_{bio}, including the additional constraint of a small lack of closure of $sDIC_{calc}$ during the annual cycle. Thus our best estimate of F_{bio}, and hence of net community production integrated over the mixed layer, is 11 ± 6 gC m^{-2} yr^{-1}.

8.2 COMPARISON WITH OTHER ESTIMATES

We now compare the estimates of carbon fluxes in the waters near Bermuda based on our model calculations with estimates based on other observations and other types of models. We focus on comparisons of the air-sea exchange and biological exchange for which other data are most readily available.

Air-sea gas exchange

The pCO_{2oc} data for Station S, indicate that the waters near Bermuda are a strong sink for CO_2, provided that a correction of 10 ppm is applied to our pCO_{2oc} data derived from DIC and alkalinity measurements (see Section 4). These pCO_{2oc} data would have to be raised uniformly by 34 ppm throughout the year to achieve a zero net air-sea flux. This difference is larger than the annual average of the air-sea difference

$$\Delta pCO_2 = pCO_{2atm} - pCO_{2oc} \tag{28}$$

equal to -23 ppm, because the gas-exchange rate tends to be higher when ΔpCO_2 is negative, especially in winter when winds are strong.

The annual net uptake of atmospheric CO_2 by ocean water at Station S, indicated by our calculations, is approximately 21 ± 6 gC m^{-2} yr^{-1} (see Subsection 8.1). This estimate can be compared with the results of recent seasonal compilation of ΔpCO_2 over the North Atlantic by Takahashi, Takahashi, and Sutherland (1995). They used a numerical interpolation scheme based upon the lateral diffusive and advective transport of surface waters to interpolate measurements made irregularly in time and space. They also calculated the monthly air-sea fluxes of CO_2 using the gas-exchange coefficients of Tans, Fung, and Takahashi (1990) and integrated them during the year to obtain annual net air-sea exchange fluxes. For the 4° latitude 5° longitude box encompassing Bermuda they found an annual ocean uptake of about $3.1 \cdot 10^{12}$ gC yr^{-1}, implying a flux of about 15 gC m^{-2} yr^{-1}, based on an area of $2^{10} \cdot 10^{11}$ m^2, within the uncertainties of our estimate. Our estimate, however, is substantially higher than that for the BATS station by Bates, Michaels, and Knap (1996, 375) (average of 7.41 ± 43 gC m^{-2} yr^{-1} for 1989, 1991, 1992, and 1993) computed from direct observations in a manner similar to our computation.

Biological exchange

The magnitude of primary and new production in the waters near Bermuda has been repeatedly debated. Several recent studies yield estimates of both primary and new production at Station S and elsewhere in the Sargasso Sea. A summary of these, and of earlier estimates of Menzel and Ryther (1960), is given in Table 7. As pointed out in Section 2, above, we will assume in this discussion that during the annual cycle net community production, new production, and the export of organic matter from the euphotic layer are equal.

Several of the studies have inferred production from oxygen measurements or from nitrogen-based ecosystem models. The associated estimates are thus sensitive to the choice of the photosynthetic quotient (PQ) and the C:N ratios in organic matter (see Section 2, above). The traditional Redfield ratios (Redfield, Ketchum, and Richards 1963) are today a topic of lively discussion, and a number of modifications have been proposed (Takahashi, Broecker, and Langer 1985, Laws 1991, Sambrotto et al. 1993, Banse 1994, Anderson and Sarmiento 1994). In Table 7 we chose the average P:N:C:O$_2$ ratios of Takahashi, Broecker, and Langer (1985): (1:16:122:-172) for the conversion of oxygen into carbon, thus a PQ of 1.41. This choice is supported by Laws (1991), who argued, based on balanced chemical equations for the production of organic matter, that the PQ of new production is about 1.4 ± 0.1. By using this PQ we have lowered the estimate of new production by Jenkins and Goldman (1985) who originally used a PQ of 1.24. The same PQ was used for the estimates of Musgrave, Chou, and Jenkins (1988) and Spitzer and Jenkins (1989). We also used the C:N ratio of Takahashi, Broecker, and Langer (1985) (122:16) for the conversion of nitrogen to carbon. This ratio has been put into considerable question by Sambrotto et al. (1993) but was recently supported in a study by Anderson and Sarmiento (1994).

In Table 7, the estimate of particulate organic carbon (POC) export flux by Lohrenz et al., (1992) using sediment traps has been obtained at a deployment depth of 150 m. We estimated the POC flux at 100 m by scaling the reported value by a factor of 1.5 based on the empirical relationship of Martin et al. (1987).

Our study for the mixed layer yields an annual decrease in concentration caused by *in situ* biological processes ($\Delta sDIC_{bio}/\Delta t$) of -0.42 gC m^{-3} yr^{-1}, (see Table 4, above). As is evident from our discussion in Subsection 8.1, this estimate is sensitive to the magnitude of gas exchange and vertical diffusive transport, but only slightly influenced by likely uncertainties in the isotopic fractionation factor, α_{org}, or in entrainment. The corresponding negative of net biological exchange flux (our estimate of net community production) is computed to be -10.7 gC m^{-2} yr^{-1} for the mixed layer (Table 5).

In order to make comparisons with other estimates of biological exchange, this flux has to be evaluated over the whole euphotic layer rather than over only the mixed layer. If our computed concentration change in the mixed layer were representative for the euphotic layer to its average depth of about 100 m (Marra et al. 1992), the annual net community production would be about 42 gC m^{-2} yr^{-1}. This estimate is almost surely too high, because net community production usually decreases with depth over the euphotic layer, reflecting a much more rapid decrease with depth in net primary production than in respiration (Longhurst and Williams 1979).

Based on four vertical radiocarbon productivity profiles measured by Marra et al. (1992) in 1987 and our determination of the seasonal cycle in mixed-layer depth, we estimate the fraction of annual primary production occurring below the mixed layer to be about 50% of the water-column integrated production. Fasham, Ducklow, and McKelvie (1990), however, estimated this value to be only 27% based on a reanalysis of the Menzel and Ryther (1960) data for 1959. Using the average of these two estimated fractions (38%) and assuming that the same fraction applies for net community production, our computed net biological flux of about 11 gC m^{-2} yr^{-1} in the mixed layer is equivalent to about 18 gC m^{-2} yr^{-1} integrated over the euphotic layer. However, we point out that we lack observations to support the assumption of a constant ratio between net community production and net primary production over the euphotic layer. This simple assumption is inconsistent with the concept of a two-layered euphotic layer as discussed by Jenkins and Goldman (1985), Small, Knauer, and Tuel (1987), and Goldman (1988), in which most of the primary production occurs in the upper mixed layer and is fueled mainly by regenerative nutrients, whereas the bulk of new production is occurring at or near the base of the euphotic layer. Thus the extrapolation of our result from the mixed layer to the whole euphotic layer must be viewed as tentative.

This extrapolated estimate of net community production is lower than, but within reasonable agreement with, estimates of the export of total organic carbon (TOC) based on POC export inferred from sediment trap data of about 14 gC m^{-2} yr^{-1} (Lohrenz et al. 1992) and based on dissolved organic carbon (DOC) export inferred from DOC cycles of about

13 gC m^{-2} yr^{-1} (Carlson, Ducklow, and Michaels 1994). Our value, however, is significantly lower than estimates of new production of about 40 gC m^{-2} yr^{-1} as based either on the seasonal cycle of oxygen (Jenkins and Goldman 1985, Musgrave, Chou, and Jenkins 1988, Spitzer and Jenkins 1989) or on modeling studies (Fasham, Ducklow, and McKelvie 1990, Fasham et al. 1993). The different estimates, however, must be cautiously compared, because they refer to different types of production, as discussed in Section 2.

The estimated flux of POC out of the euphotic layer as inferred from sediment traps is associated with large uncertainties (Michaels et al. 1994b, Buesseler et al. 1994) because of sampling problems with these traps, which include hydrodynamic biases (under- and over-collection), interference with swimmers (zooplankton that actively enter the traps), and solubilization of particulate matter in the traps. Although they do not state error margins, the estimate of DOC export from the euphotic layer by convective mixing given by Carlson, Ducklow, and Michaels (1994) is probably also quite uncertain.

The export of particulate nitrogen below the euphotic layer between 1986 and 1988 was estimated by Altabet (1989a) to be about 0.2 mol N m^{-2} yr^{-1}. He also estimated that 0.13 mol N m^{-2} yr^{-1} is additionally exported by downward mixing of suspended particulate nitrogen. These values are equivalent to an export of about 18 gC m^{-2} yr^{-1} and 12 gC m^{-2} yr^{-1}, respectively, when the Takahashi, Broecker, and Langer (1985) C:N ratio of 122:16 is applied. However, his estimate of the downward mixing of particulate nitrogen is based on relatively high vertical diffusion coefficients and may therefore represent an overestimate.

Given the uncertainties, our estimate of annual net community production for the euphotic layer is consistent with these estimates of export production.

Based on a nitrogen cycle model of the plankton dynamics in the mixed layer at Station S, Fasham, Ducklow, and McKelvie (1990) estimated the annual new production in the mixed layer to be about 0.4 mol N m^{-2} yr^{-1}. They adjusted their model to have an annual net primary production in the mixed layer of 45 gC m^{-2} yr^{-1} in accordance with the data of Menzel and Ryther (1960). Assuming a Redfield C:N ratio of 122:16, this converts to a carbon new production of 37 gC m^{-2} yr^{-1} in the mixed layer, more than three times our mixed-layer net community production of 11 gC m^{-2} yr^{-1}. Fasham, Ducklow, and McKelvie (1990) extrapolated the mixed-layer value to the whole euphotic zone by using their estimate of the fraction of net primary production occurring below the mixed layer (27%, see above) and assuming a constant f-ratio over the euphotic layer. This resulted in an estimated annual new production over the euphotic layer of about 47 gC m^{-2} yr^{-1} in good accordance with the results based on oxygen measurements (Jenkins and Goldman 1985, Musgrave, Chou, and Jenkins 1988, Spitzer and Jenkins 1989).

The ecosystem model of Fasham, Ducklow, and McKelvie (1990) was later included in a seasonally forced three-dimensional general circulation model of the North Atlantic Ocean (Fasham et al 1993, Sarmiento et al. 1993). In this model, the same parameters for the ecosystem model were used as previously, but the results pertain to the whole euphotic layer, assumed to be 113 m deep. New production was found to be between 0.38 and 0.57 mol N m^{-2} yr^{-1} (thus 35 to 52 gC m^{-2} yr^{-1}).

A possible cause for at least part of the discrepancy between our estimate and the model estimates of Fasham and his coauthors may be caused by their overestimation of nitrate transport into the mixed or euphotic layer. With respect to their earlier model (Fasham, Ducklow, and McKelvie 1990), an overestimate may have been caused by their assuming too high nitrate concentrations below the mixed layer, resulting in high entrainment fluxes, whereas in their later model (Fasham et al. 1993) this may have been a result of an artifact of the physical model causing too high advective input of nitrate during the winter months (*loc. cit.*, 387).

8.3 COMPARISON WITH TRACER TRANSPORT MODEL PREDICTIONS

It is possible to put our estimates of fluxes of carbon in the waters near Bermuda into the perspective of the large-scale circulation of the oceans by comparing them with the predictions of the three-dimensional tracer model of Bacastow and Maier-Reimer (1991), a model that we have already used to provide estimates of the vertical gradient of the $^{13}C/^{12}C$ ratio of DIC (see Subsection 6.3, above). This nonseasonal model with a uniform 115 m deep surface layer is described in more detail in Appendix B. In order to compare our fluxes that pertain to the mixed layer with the fluxes of the surface box of their model near Bermuda, we have converted our estimates to those for the euphotic layer, which is approximately the same depth at Bermuda as their model surface layer.

We have already estimated net community production for the euphotic layer to be about 18 gC m^{-2} yr^{-1} by extrapolating from our estimate for the mixed layer on the basis of net primary production data (see Subsection 8.2 above). Let us now assume that the diffusive flux, F_{diff}, during the annual cycle into the euphotic layer is the same as that into the mixed layer. Thus we deduce, as a first approximation, that the vertical diffusive flux adds 15 gC m^{-2} yr^{-1} of DIC to the euphotic layer. We neglect entrainment, because it is small in our model, and because the base of the euphotic layer is fixed. The air-sea flux, F_{ex}, adds 21 gC m^{-2} yr^{-1} to the water column, independent of the depth of the layer considered; hence it applies directly to the euphotic layer. These fluxes do not even approximately sum to zero as they should for a fixed layer at steady state; rather, they imply that DIC in the euphotic layer near Bermuda increases at the rate of 18 gC m^{-2} yr^{-1}.

In seeking an explanation, we turn to the results of the above cited three-dimensional transport model. This model predicts fluxes for biological exchange and vertical mixing in their surface layer of 115 m depth that agree well with our estimates, a loss of 17 gC m^{-2} yr^{-1} by conversion of DIC to organic carbon *in situ*, and a gain of nearly the same amount by vertical transport. The model, however, predicts a much lower uptake of CO_2 by the air-sea flux, only 7 gC m^{-2} yr^{-1}, and a nearly equal offsetting flux by lateral transport coupled with downwelling thus achieving a steady state (See Figure 17). Evidently, the imbalance in our calculations arises mainly because of a larger uptake of CO_2 by air-sea exchange, with no compensating lateral transport.

The air-sea flux predicted by the three-dimensional model of Bacastow and Maier-Reimer (1991) is mainly caused by the Suess effect (4 gC m^{-2} yr^{-1}) and by the temperature dependency of the CO_2 solubility (2 gC m^{-2} yr^{-1}). Thus, the model shows no evidence of a "North Atlantic Sink Component," which is predicted by Broecker and Peng (1992). Their prediction is based on computing a southward transport DIC from the Northern Hemisphere into the southern via deep water and which Keeling, Piper, and Heimann (1989) invoked to help explain the observed global-scale north-south gradient in atmospheric CO_2.

If we accept the three-dimensional model prediction that about 7 gC m^{-2} yr^{-1} are exported from the euphotic layer by lateral transport at Station S, the annual imbalance in our estimates of the fluxes in and out of the euphotic layer there reduces to 11 gC m^{-2} yr^{-1}. This value is still considerable but within the margin of error of the sum of our individual estimates of fluxes. A possible cause for part of this remaining imbalance is that the air-sea flux is less than 21 gC m^{-2} yr^{-1}, for example as suggested by the estimate of Takahashi, Takahashi, and Sutherland (1995) of 15 gC m^{-2} yr^{-1} for the waters near Bermuda. The remaining imbalance is explained should we have considerably overestimated the vertical diffusive transport in our standard case. As shown in our sensitivity tests, our mixed-layer model is rather insensitive to whether the vertical transport into the mixed layer is dominated by entrainment or diffusion, as long as their sum is more or less constant. If we assume that vertical diffusion in the upper ocean is small throughout the year, consistent with microstructure measurements and tracer experiments (see Subsection 6.3), and if entrainment is then increased in the model to preserve approximately the same vertical transport as our standard case, the annual budget for the mixed layer is not significantly affected, as pointed out by a sensitivity test (see Subsection 8.1). The vertical transport of carbon into the euphotic layer, however, is reduced because entrainment transports carbon into that layer only when the mixed layer is extending below it. Thus our calculations of air-sea exchange and vertical transport for the euphotic layer are too uncertain to distinguish our budget estimates for this layer from those of the fixed surface layer of the three-dimensional transport model.

In either case there is indication that lateral transport, missing from our one dimensional calculations, together with biological processes remove DIC from the surface mixed layer on the annual timescale. The question arises also, whether lateral advection may have a significant impact on the seasonal carbon cycle at Station S. Michaels et al. (1994b) argued that in order to account for the discrepancy in their carbon balance figures for the top 150 m at station BATS from April through December, sediment traps have severely undercollected the downward flux of POC and/or horizontal advection has contributed significantly to the seasonal carbon cycle near Bermuda. At present, we can only speculate because we lack more information about the circulation patterns and the spatial distribution of DIC and $^{13}\delta$ in the vicinity of Bermuda. In order to explain the seasonal drawdown of carbon resulting only from physical processes, advection would need to have a very negative isotopic signature, similar to that of organic carbon. Sparse information about the spatial distribution of $^{13}\delta$ in the North Atlantic (unpublished data by C. D. Keeling and coworkers) does not show strong gradients in $^{13}\delta$ required to generate such an isotopic signature. We therefore tentatively conclude that advective transport plays a significant role for the annual carbon balance at Station S near Bermuda, but only a minor role in determining the seasonal signal.

9. SUMMARY AND CONCLUSIONS

The seasonal cycle of DIC in the surface waters near Bermuda is only partially a result of biological processes. To determine the rate at which DIC is converted to organic carbon via photosynthesis and is regenerated by oxidation of organic matter, observed changes in DIC must first be corrected to account for transport processes that alter the DIC concentration, including air-sea exchange, vertical diffusion, and entrainment.

We have estimated in situ biological exchange (defined as positive when DIC increases) in two ways; first from changes in the $^{13}C/^{12}C$ ratio of DIC that are caused by a large difference in isotopic ratio between photosynthetic carbon and DIC, and second by subtracting model calculated rates of change in DIC concentration from observed rates. The two approaches give similar results after correcting the $^{13}C/^{12}C$ signal of DIC for isotopic fractionation associated with the same processes that influence the DIC concentration, confirming the consistency of our approach.

The resulting net biological exchange flux by either model is approximately equivalent to the negative of net community production. Over the year, if interannual variations are negligible, net community production is equal to new production, and also to the export of organic matter out of the water layer under consideration.

We made use of empirical relationships to estimate the contributions of air-sea exchange, vertical mixing and entrainment, and incorporated them into a seasonal box model of the mixed layer. Given the uncertainties in the calculations, the agreement in the mixed layer between the modeled and observed seasonal variations of DIC normalized to constant salinity (sDIC) is satisfactory, although modeled sDIC failed to close exactly over the annual cycle. We deduced that air-sea exchange and vertical transport each add about 21 gC m^{-2} yr^{-1} of CO_2 to the mixed layer. We infer that net community production removes 11 gC m^{-2} yr^{-1} over the year. This is a result of a net removal of 19 gC m^{-2} during the shoaling period from February to July and a net gain of 8 gC m^{-2} during the deepening period. Sensitivity analysis shows that our estimate of net community production depends strongly on the magnitude of the isotopic fractionation during air-sea exchange, and to a lesser, but significant, degree on variations in the magnitude of the vertical transport processes. From separate computations of uncertainties of about 30% for air-sea exchange and about 50% for vertical transport, we conclude that our approximation of net community production has an uncertainty on the order of 50%. Our estimate of air-sea exchange of 21 gC m^{-2} yr^{-1} is about midway between the estimate of 30 gC m^{-2} based on atmospheric CO_2 gradients (Keeling, Piper, and Heimann 1989b), and of 15 gC m^{-2} based on regional North Atlantic pCO$_{2oc}$ data (Tans, Fung, and Takahashi 1990), but substantially higher than set forth by Bates (1996) for the nearby BATS station. We cannot resolve these discrepancies without more information. The rather large errors in our calculation of air-sea exchange could be reduced considerably by resolving the discrepancy between our calculation of oceanic CO_2 partial pressure using thermodynamic constants and the direct measurement of partial pressure using gas equilibrators, a task that is under way in our laboratory.

To compare our estimate of net community production for the mixed layer with several previous calculations of new production and of downward export of particulate and previous calculations of new production and of downward export of particulate and dissolved organic matter, we estimated net community production for the euphotic layer. Based on limited evidence from the vertical distribution of net primary production and the assumption of a constant ratio of net community production to net primary production within the euphotic layer, we extrapolated our estimate for the mixed layer to the whole euphotic zone. We obtained a tentative estimate of 18 gC m^{-2} yr^{-1}, which is on the low side of estimates based on net downward export of total organic carbon, and of particulate nitrogen measured by sediment traps and water-column inventories, but which agrees within the uncertainties of these estimates and of ours. Our euphotic net community production, however, is significantly lower than estimates of new production based on oxygen and noble gas measurements and on modeling studies.

Our calculated carbon fluxes for the mixed-layer sum to a net gain of 28 gC m^{-2} yr^{-1} for the mixed layer, in good agreement with the value of 27 gC m^{-2} yr^{-1} derived from observations of the seasonal cycles of the mixed-layer depth and of DIC. This apparent gain is balanced by an equal loss term associated with the variable mixed-layer depth, because the mixed layer shoals when the concentration of DIC, on average, is higher than when it deepens. To establish the carbon balance for the whole euphotic layer, assumed to have a constant depth of 100 m, we compared our findings with predictions of the three-dimensional ocean tracer transport model of Bacastow and Maier-Reimer (1991). This shows lateral transport, missing in our calculations. That model predicts that this transport removes approximately 7 gC m^{-2} yr^{-1} from the upper 115 m near Bermuda supporting the hypothesis of Michaels et al. (1994b) that horizontal advection may contribute significantly to the carbon cycle near Bermuda. Including this transport into our budget for the whole euphotic layer leaves an imbalance of 11 gC m^{-2} yr^{-1} which, however, disappears, if our estimate of the gain in CO_2 from air-sea exchange is too high, and/or the dominant vertical transport process is entrainment and not vertical diffusion.

Compared to relying on short-term discrete measurements of organic production (e.g. by incubation using radiocarbon) our integrative geochemical approach is not sensitive to short-term variations in biological activity. Therefore our estimate of net community production of 11 gC m^{-2} yr^{-1} in the mixed layer should not be biased on the low side by suspected short phytoplankton blooms. Our method is also not dependent on any assumption about the magnitude and variability of the photosynthetic quotients or of the Redfield ratios for carbon, oxygen, and nutrients. The strength of our analysis comes from using six years of sufficiently closely spaced data to establish reliably the average seasonal carbon cycle at Station S, or at least the most reliable seasonal cycle yet obtained anywhere in the world oceans. Our model predictions are, however, restricted to the surface mixed layer. Adequate measurements of the seasonal cycle of DIC concentration and $^{13}C/^{12}C$ ratio below this layer would allow the carbon balances for the whole euphotic layer to be established and would reduce errors considerably in calculating vertical transport, and thus net community production.

In our study we did not address the problem of how the phytoplankton in the mixed layer meet the nutrient requirements for supporting our estimated net community production. The observed nutrient concentrations in the mixed layer and estimates of nutrient transport into the mixed layer are too low to support our estimate of net community production if the standard Redfield ratios are applied. We confess that we still do not understand the nutrient cycles in these oligotrophic regions and possible adaptations of the plankton communities to these environments.

Chemical measurements near Bermuda, which still continue, offer the possibility to look not only at the seasonal carbon cycle but also at longer-term variability. Our analyses of the average seasonal cycle should help in interpreting such records, including the influence of the build-up of industrial CO_2 and climatic variability. Our results for Station S in the Sargasso Sea, although not shown to apply to large regions of the world ocean, are perhaps sufficiently representative of the North Atlantic central gyre to provide a first direct estimate of the average seasonal cycle of carbon in this broad and important subtropical region.

REFERENCES

Anderson, Laurence A., and Jorge L. Sarmiento. 1994. Redfield ratios of remineralization determined by nutrient data analysis. Global Biogeochemical Cycles 8(1):65-80.

Altabet, Mark A. 1989a. Particulate new nitrogen fluxes in the Sargasso Sea. Journal of Geophysical Research 94(C9):12,771-12,779.

Altabet, Mark A. 1989b. A time-series study of the vertical structure of nitrogen and particle dynamics in the Sargasso Sea. Limnology and Oceanography 34(7):1185-1201.

Bacastow, Robert. 1981. Numerical evaluation of the evasion factor. In SCOPE 16: Carbon Cycle Modelling, edited by Bert Bolin. John Wiley and Sons, New York:95-101.

Bacastow, R., and E. Maier-Reimer. 1990. Ocean-circulation model of the carbon cycle. Climate Dynamics 4:95-125.

Bacastow, R., and E. Maier-Reimer. 1991. Dissolved organic carbon in modeling oceanic new production. Global Biogeochemical Cycles 5(1):71-85.

Bacastow, Robert B., Charles D. Keeling, Timothy J. Lueker, Martin Wahlen, and Willem G. Mook. 1996. The ^{13}C Suess effect in the world surface oceans and its implications for oceanic uptake of CO_2: Analysis of observations at Bermuda. Global Biogeochemical Cycles 10(2):335-346.

Baines, Stephen B., and Michael L. Pace. 1991. The production of dissolved organic matter by phytoplankton and its importance to bacteria: Patterns across marine and freshwater systems. Limnology and Oceanography 36(6):1078-1090.

Banse, K. 1994. Uptake of inorganic carbon and nitrate by marine plankton and the Redfield ratio. Global Biogeochemical Cycles 8(1):81-84.

Bates, Nicholas R., Anthony F. Michaels, and Anthony H. Knap. 1996. Seasonal and interannual variability of oceanic carbon dioxide species at the U.S. JGOFS Bermuda Atlantic Time-series Study (BATS) site. Deep-Sea Research II 43(2-3):347-383.

Bender, Michael, Karen Grande, Kenneth Johnson, John Marra, Peter J. LeB. Williams, John Sieburth, Michael Pilson, Chris Langdon, Gary Hitchcock, Joseph Orchardo, Carleton Hunt, Percy Donaghay, and Kristina Heinemann. 1987. A comparison of four methods for determining planktonic community production. Limnology and Oceanography 32(5):1085-1098.

Broecker, Wallace Smith, and Tsung-Hung Peng. 1982. Tracers in the Sea. Lamont-Doherty Geological Observatory, Columbia University, Palisades, New York. 690 p.

Broecker, Wallace S., and Tsung-Hung Peng. 1992. Interhemispheric transport of carbon dioxide by ocean circulation. Nature 356:587-589.

Brown, Christopher W., and James A. Yoder. 1994. Coccolithophorid blooms in the global ocean. Journal of Geophysical Research 99(C4):7467-7482.

Buesseler, Ken O., Anthony F. Michaels, David A. Siegel, and Anthony H. Knap. 1994. A three dimensional time-dependent approach to calibrating sediment trap fluxes. Global Biogeochemical Cycles 8(2):179-193.

Burris, J. E. 1981. Effects of oxygen and inorganic carbon concentrations on the photosynthetic quotients of marine algae. Marine Biology 65:215-219.

Carlson, Craig A., Hugh W. Ducklow, and Anthony F. Michaels. 1994. Annual flux of dissolved organic carbon from the euphotic zone in the northwestern Sargasso Sea. Nature 371:405-408.

Chipman, David W., John Marra, and Taro Takahashi. 1993. Primary production at 47°N and 20°W in the North Atlantic Ocean: a comparison between the ^{14}C incubation method and the mixed layer carbon budget. Deep-Sea Research II 40(1/2):151-169.

Deacon, E. L. 1977. Gas transfer to and across an air-water interface. Tellus 29:363-374.

Degens, E. T., R. R. L. Guillard, W. M. Sackett, and J. A. Hellebust. 1968. Metabolic fractionation of carbon isotopes in marine plankton - I. Temperature and respiration experiments. Deep-Sea Research 15:1-9.

Degens, E. T. 1969. Biogeochemistry of stable carbon isotopes. In: Organic Geochemistry: Methods and Results, edited by G. Eglinton and M. T. J. Murphy. Springer-Verlag, Berlin:304-329.

Denman, K. L., and A. E. Gargett. 1983. Time and space scales of vertical mixing and advection of phytoplankton in the upper ocean. Limnology and Oceanography 28(5):801-815.

Deuser, W. G., E. T. Degens, and R. R. L. Guillard. 1968. Carbon isotope relationships between plankton and sea water. Geochimica et Cosmochimica Acta 32:657-660.

Dickson, A. G., and F. J. Millero. 1987. A comparison of the equilibrium constants for the dissociation of carbonic acid in seawater media. Deep-Sea Research 34(10): 1733-1743.

Dillon, Thomas M., and Douglas R. Caldwell. 1980. The Batchelor spectrum and dissipation in the upper ocean. Journal of Geophysical Research 85(C4):1910-1916.

Druffel, Ellen R. M., Peter M. Williams, James E. Bauer, and John R. Ertel. 1992. Cycling of dissolved and particulate organic matter in the open ocean. Journal of Geophysical Research 97(C10):15,639-15,659.

Dugdale, R. C., and J. J. Goering. 1967. Uptake of new and regenerated forms of nitrogen in primary productivity. Limnology and Oceanography 32:196-206.

Emerson, Steven, Paul Quay, Charles Stump, David Wilbur, and Rebecca Schudlich. 1993. Determining primary production from the mesoscale oxygen field. In: Measurement of Primary Production from the Molecular to the Global Scale: Proceedings of a Symposium held in La Rochelle, 21-24 April 1992, edited by William K. W. Li and Serge Y. Maestrini. ICES Marine Science Symposia 197:196-206.

Eppley, Richard W., and Bruce J. Peterson. 1979. Particulate organic matter flux and planktonic new production in the deep ocean. Nature 282:677-680.

Fasham, M. J. R., H. W. Ducklow, and S. M. McKelvie. 1990. A nitrogen-based model of plankton dynamics in the oceanic mixed layer. Journal of Marine Research 48(3):591-639.

Fasham, M. J. R., J. L. Sarmiento, R. D. Slater, H. W. Ducklow, and R. Williams. 1993. Ecosystem behavior at Bermuda Station "S" and Ocean Weather Station "India": A general circulation model and observational analysis. Global Biogeochemical Cycles 7(2):379-415.

Friedli, H., H. Lötscher, H. Oeschger, U. Siegenthaler, and B. Stauffer. 1986. Ice core record of the $^{13}C/^{12}C$ ratio of atmospheric CO_2 in the past two centuries. Nature 324: 237-238.

Goericke, Ralf, and Brian Fry. 1994. Variations of marine plankton $\delta^{13}C$ with latitude, temperature, and dissolved CO_2 in the world ocean. Global Biogeochemical Cycles 8(1):85-90.

Goldman, Joel C. 1988. Spatial and temporal discontinuities of biological processes in pelagic surface waters. In: Toward a Theory on Biological-Physical Interactions in the World Ocean, edited by B. J. Rothschild. Kluwer Academic Publishers, Amsterdam:273-296.

Gruber, N., C. D. Keeling, and T. F. Stocker. 1998. Carbon-13 constraints on the seasonal inorganic carbon budget at the BATS site in the northwestern Sargasso Sea. Deep-Sea Research, Part I, 45, 673-717.

Hayward, Thomas L. 1994. The shallow oxygen maximum layer and primary production. Deep-Sea Research I 41(3):559-574.

Heimann, Martin, and Charles D. Keeling. 1989. A three-dimensional model of atmospheric CO_2 transport based on observed winds: 2. Model description and simulated tracer experiments. In: Aspects of Climate Variability in the Pacific and the Western Americas, Geophysical Monograph 55, edited by David H. Peterson. American Geophysical Union, Washington, D.C.:237-275.

Heimann, Martin, Charles D. Keeling, and Compton J. Tucker. 1989. A three dimensional model of atmospheric CO_2 transport based on observed winds: 3. Seasonal cycle and synoptic time scale variations. In: Aspects of Climate Variability in the Pacific and the Western Americas, Geophysical Monograph 55, edited by David H. Peterson. American Geophysical Union, Washington, D.C.:277-303.

Heimann, Martin, and Patrick Monfray. 1989. Spatial and temporal variation of the gas exchange coefficient for CO_2: 1. Data analysis and global validation. Report No. 31, Max-Planck-Institut für Meteorologie, Hamburg, Germany. 29 p.

Isemer, Hans-Jörg, and Lutz Hasse. 1985. The Bunker Climate Atlas of the North Atlantic Ocean Volume 1: Observations. Springer-Verlag, Berlin. 218 p.

Jenkins, W. J., and J. C. Goldman. 1985. Seasonal oxygen cycling and primary production in the Sargasso Sea. Journal of Marine Research 43(2):465-491.

Jenkins, W. J. 1988. Nitrate flux into the euphotic zone near Bermuda. Nature 331:521-523.

Karl, David M., and George A. Knauer. 1989. Swimmers: a recapitulation of the problem and a potential solution. Oceanography Magazine 2(1):32-35.

Karl, David M., Bronte D. Tilbrook, and Georgia Tien. 1991. Seasonal coupling of organic matter production and particle flux in the western Bransfield Strait, Antarctica. Deep-Sea Research 38(8/9):1097-1126.

Karl, David M., and Roger Lukas. 1996. The Hawaii Ocean Time-series (HOT) program: Background, rational and field implementation. Deep-Sea Research II 43(2-3):129-156.

Keeling, Charles D., R. B. Bacastow, A. F. Carter, S. C. Piper, Timothy P. Whorf, Martin Heimann, Willem G. Mook, and Hans Roeloffzen. 1989. A three-dimensional model of atmospheric CO_2 transport based on observed winds: 1. Analysis of observational data. In: Aspects of Climate Variability in the Pacific and the Western Americas, Geophysical Monograph 55, edited by David H. Peterson. American Geophysical Union, Washington, D.C.:165-236.

Keeling, Charles D., Stephen C. Piper, and Martin Heimann. 1989. A three-dimensional model of atmospheric CO_2 transport based on observed winds: 4. Mean annual gradients and interannual variations. In: Aspects of Climate Variability in the Pacific and the Western Americas, Geophysical Monograph 55, edited by David H. Peterson. American Geophysical Union, Washington, D.C.:305-363.

Keeling, Charles D. 1993. Lecture 2: Surface ocean CO_2. In: The Global Carbon Cycle, edited by Martin Heimann. Springer-Verlag, Berlin:413-429.

Keeling, Ralph F. 1993. On the role of large bubbles in air-sea gas exchange and supersaturation in the ocean. Journal of Marine Research 51(2):237-271.

Kroopnick, P. M. 1985. The distribution of ^{13}C of ΣCO_2 in the world oceans. Deep-Sea Research 32(1):57-84.

Kurz, Katharina Dorothea. 1993. Zur saisonalen Variation des ozeanischen Kohlendioxidpartialdrucks. PhD. Thesis, University of Hamburg. Max-Planck-Institut für Meteorologie, Hamburg, Germany: 115 p.

Large, W. G., J. C. McWilliams, and P. P. Niiler. 1986. Upper ocean thermal response to strong autumnal forcing of the northeast Pacific. Journal of Physical Oceanography 16:1524-1550.

Laws, Edward A. 1991. Photosynthetic quotients, new production and net community production in the open ocean. Deep-Sea Research 38(1):143-167.

Ledwell, James R., Andrew J. Watson, and Clifford S. Law. 1993. Evidence for slow mixing across the pycnocline from an open-ocean tracer-release experiment. Nature 364: 701-703.

Levitus, Sydney. 1982. Climatological Atlas of the World Ocean. NOAA Professional Paper 13, U.S. Government Printing Office, Washington, D.C. 173 p.

Lewis, Marlon R., W. Glen Harrison, Neil S. Oakey, David Hebert, and Trevor Platt. 1986. Vertical nitrate fluxes in the oligotrophic ocean. Science 234:870-873.

Liss, Peter S., and Liliane Merlivat. 1986. Air-sea gas exchange rates: introduction and synthesis. In: The Role of Air-Sea Exchange in Geochemical Cycling, edited by Patrick Buat-Ménard. D. Reidel Publishing Company, Boston:113-127.

Lohrenz, Steven E., George A. Knauer, Vernon L. Asper, Merritt Tuel, Anthony F. Michaels, and Anthony H. Knap. 1992. Seasonal variability in primary production and particle flux in the northwestern Sargasso Sea: U.S. JGOFS Bermuda Atlantic Time-series Study. Deep-Sea Research 39(7/8):1373-1391.

Longhurst, Alan R. 1991. Role of the marine biosphere in the global carbon cycle. Limnology and Oceanography 36(8):1507-1526.

Longhurst, Alan, and Robert Williams. 1979. Materials for plankton modelling: Vertical distribution of Atlantic zooplankton in summer. Journal of Plankton Research 1(1): 1-28.

Malone, Thomas C., Sharon E. Pike, and Daniel J. Conley. 1993. Transient variations in phytoplankton productivity at the JGOFS Bermuda time series station. Deep-Sea Research I 40(5):903-924.

Marra, John, T. Dickey, W. S. Chamberlin, C. Ho, T. Granata, D. A. Kiefer, C. Langdon, R. Smith, K. Baker, R. Bidigare, and M. Hamilton. 1992. Estimation of seasonal primary production from moored optical sensors in the Sargasso Sea. Journal of Geophysical Research 97(C5):7399-7412.

Martin, John H., George A. Knauer, David M. Karl, and William W. Broenkow. 1987. VERTEX: carbon cycling in the northeast Pacific. Deep-Sea Research 34(2): 267-285.

Menzel, D. W., and J. H. Ryther. 1960. The annual cycle of primary production in the Sargasso Sea off Bermuda. Deep-Sea Research 6:351-367.

Menzel, D. W., and J. H. Ryther. 1961. Annual variations in primary production of the Sargasso Sea off Bermuda. Deep-Sea Research 7:282-288.

Michaels, Anthony F., Anthony H. Knap, and John W. H. Dacey. 1992. The U.S. JGOFS Bermuda Atlantic Time-series Study: Towards an understanding of the temporal and spatial scales of ocean biogeochemistry. Washington, D.C., Marine Technical Society, MTS '92 Conference Proceedings 2:535-541.

Michaels, Anthony F., Anthony H. Knap, Rachael L. Dow, Kjell Gundersen, Rodney J. Johnson, Jens Sorensen, Ann Close, George A. Knauer, Steven E. Lohrenz, Vernon A. Asper, Merritt Tuel, and Robert Bidigare. 1994a. Seasonal patterns of ocean biogeochemistry at the U.S. JGOFS Bermuda Atlantic Time-series Study site. Deep-Sea Research I 41(7):1013-1038.

Michaels, Anthony F., Nicholas R. Bates, Ken O. Buesseler, Craig A. Carlson, and Anthony H. Knap. 1994b. Carbon-cycle imbalances in the Sargasso Sea. Nature 372:537-540.

Michaels, Anthony F., and Anthony H. Knap. 1996. Overview of the U.S. JGOFS Bermuda Atlantic Time-series Study and the Hydrostation S program. Deep-Sea Research II 43(2-3):157-198.

Millero, Frank J., Chen-Tung Chen, Alvin Bradshaw, and Karl Schleicher. 1980. A new high pressure equation of state for seawater. Deep-Sea Research 27A:255-264.

Millero, Frank J., Robert H. Byrne, Rik Wanninkhof, Richard Feely, Tonya Clayton, Paulette Murphy, and Marilyn F. Lamb. 1993. The internal consistency of CO_2 measurements in the equatorial Pacific. Marine Chemistry 44:269-280.

Mook, W. G., J. C. Bommerson, and W. H. Staverman. 1974. Carbon isotope fractionation between dissolved bicarbonate and gaseous carbon dioxide. Earth and Planetary Science Letters 22:169-176.

Mook, Willem G., Marjan Koopmans, Alane F. Carter, and Charles D. Keeling. 1983. Seasonal, latitudinal, and secular variations in the abundance and isotopic ratios of atmospheric carbon dioxide. 1. Results from land stations. Journal of Geophysical Research 88(C15):10,915-10,933.

Morel, André. 1988. Optical modeling of the upper ocean in relation to its biogenous matter content (case I waters). Journal of Geophysical Research 93(C9):10,749-10,768.

Musgrave, David L., James Chou, and William J. Jenkins. 1988. Application of a model of upper-ocean physics for studying seasonal cycles of oxygen. Journal of Geophysical Research 93(C12):15,679-15,700.

Najjar, Raymond G., Jorge L. Sarmiento, and J. R. Toggweiler. 1992. Downward transport and fate of organic matter in the ocean: Simulations with a General Circulation Model. Global Biogeochemical Cycles 6(1):45-76.

O'Leary, Marion H. 1981. Carbon isotope fractionation in plants. Phytochemistry 20(4):553-567.

Oudot, C. 1989. O_2 and CO_2 balances approach for estimating biological production in the mixed layer of the tropical Atlantic Ocean (Guinea Dome area). Journal of Marine Research 47(2):385-409.

Pickard, George L., and William J. Emery. 1982. Descriptive physical oceanography: An introduction. 4th enlarged edition. Pergamon Press, New York. 249p.

Phillips, O. M. 1977. Entrainment. In: Modelling and Prediction of the Upper Layers of the Ocean, edited by E. B. Kraus. Pergamon Press, New York:92-101.

Platt, Trevor, William G. Harrison, Marlon R. Lewis, William K. W. Li, Shubha Sathyendranath, Ralph E. Smith, and Alain F. Vezina. 1989. Biological production of the oceans: the case for a consensus. Marine Ecology Progress Series 52:77-88.

Platt, T., and W. G. Harrison. 1985. Biogenic fluxes of carbon and oxygen in the ocean. Nature 318:55-58.

Quay, P. D., B. Tilbrook, and C. S. Wong. 1992. Oceanic uptake of fossil fuel CO_2: Carbon-13 evidence. Science 256:74-79.

Rau, Greg H., Taro Takahashi, and David J. Des Marais. 1989. Latitudinal variations in plankton $\delta^{13}C$: implications for CO_2 and productivity in past oceans. Nature 341: 516-518.

Redfield, A. C, B. H. Ketchum, and F. A. Richards. 1963. The influence of organisms on the composition of sea-water. In: The Sea: Ideas and Observations on Progress in the Study of the Seas, volume 2, edited by M. N. Hill. Interscience, New York:26-77.

Robertson, Jane E., and Andrew J. Watson. 1993. Estimation of primary production by observation of changes in the mesoscale carbon dioxide field. In: Measurement of Primary Production from the Molecular to the Global Scale: Proceedings of a Symposium held in La Rochelle, 21-24 April 1992, edited by William K. W. Li and Serge Y. Maestrini. ICES Marine Science Symposia 197:207-214.

Robertson, J. E., C. Robinson, D. R. Turner, P. Holligan, A. J. Watson, P. Boyd, E. Fernandez, and M. Finch. 1994. The impact of a coccolithophore bloom on oceanic carbon uptake in the northeast Atlantic during summer 1991. Deep-Sea Research I 41(2):297-314.

Sambrotto, Raymond N., Graham Savidge, Carol Robinson, Philip Boyd, Taro Takahashi, David M. Karl, Chris Langdon, David Chipman, John Marra, and Louis Codispoti. 1993. Elevated consumption of carbon relative to nitrogen in the surface ocean. Nature 363:248-250.

Sarmiento, J. L., and K. Bryan. 1982. An ocean transport model for the North Atlantic. Journal of Geophysical Research 87(C1):394-408.

Sarmiento, J. L., R. D. Slater, M. J. R. Fasham, H. W. Ducklow, J. R. Toggweiler, and G. T. Evans. 1993. A seasonal three-dimensional ecosystem model of nitrogen cycling in the North Atlantic euphotic zone. Global Biogeochemical Cycles 7(2):417-450.

SCOPE. 1981. Standardization of notations and procedures. In: SCOPE 16: Carbon Cycle Modelling, edited by Bert Bolin. John Wiley and Sons, New York:81-85.

Siegel, D. A., R. Iturriaga, R. R. Bidigare, R. C. Smith, H. Pak, T. D. Dickey, J. Marra, and K. S. Baker. 1990. Meridional variations of the springtime phytoplankton community in the Sargasso Sea. Journal of Marine Research 48(2):379-412.

Small, Lawrence F., George A. Knauer, and Merritt D. Tuel. 1987. The role of sinking fecal pellets in stratified euphotic zones. Deep-Sea Research 34(10):1705-1712.

Spitzer, William S., and William J. Jenkins. 1989. Rates of vertical mixing, gas exchange and new production: Estimates from seasonal gas cycles in the upper ocean near Bermuda. Journal of Marine Research 47(1):169-196.

Sverdrup, H. U. 1953. On conditions for the vernal blooming of phytoplankton. International Council for the Exploration of the Sea: Journal du Conseil 18:287-295.

Takahashi, Taro. 1989. Only half as much CO_2 as expected from industrial emissions is accumulating in the atmosphere. Could the oceans be the storehouse for the missing gas? Oceanus 32(2):22-29.

Takahashi, Taro, Wallace S. Broecker, and Arnold E. Bainbridge. 1981. The alkalinity and total carbon dioxide concentration in the world oceans. In SCOPE 16: Carbon Cycle Modelling, edited by Bert Bolin. John Wiley and Sons, New York:271-286.

Takahashi, Taro, Wallace S. Broecker, and Sara Langer. 1985. Redfield ratio based on chemical data from isopycnal surfaces. Journal of Geophysical Research 90(C4):6907-6924.

Takahashi, Taro, Timothy T. Takahashi, and Stewart C. Sutherland. 1995. An assessment of the role of the North Atlantic as a CO_2 sink. Philosophical Transactions of the Royal Society of London B 348:143-152.

Tans, Pieter P. 1981. $^{13}C/^{12}C$ of industrial CO_2. In: SCOPE 16: Carbon Cycle Modelling, edited by Bert Bolin. John Wiley and Sons, New York:127-129.

Tans, Pieter P., Inez Y. Fung, and Taro Takahashi. 1990. Observational constraints on the global atmospheric CO_2 budget. Science 247:1431-1438.

Tans, Pieter P., Joseph A. Berry, and Ralph F. Keeling. 1993. Oceanic $^{13}C/^{12}C$ observations: A new window on ocean CO_2 uptake. Global Biogeochemical Cycles 7(2):353-368.

Thomas, F., V. Garcon, and J.-F. Minster. 1990. Modelling the seasonal cycle of dissolved oxygen in the upper ocean at Ocean Weather Station P. Deep-Sea Research 37(3):463-491.

Thomas, F., J. F. Minster, P. Gaspar, and Y. Gregoris. 1993. Comparing the behaviour of two ocean surface models in simulating dissolved O_2 concentration at O.W.S.P. Deep-Sea Research I 40(2):395-408.

Toggweiler, J. R. 1994. Vanishing in Bermuda. Nature 372:505-506.

Valiela, Ivan. 1984. Marine Ecological Processes. Springer-Verlag, Berlin. 546 p.

Villareal, Tracy A., Mark A. Altabet, and Karen Culver-Rymsza. 1993. Nitrogen transport by vertically migrating diatom mats in the North Pacific Ocean. Nature 363:709-712.

Volk, Tyler, and Martin I. Hoffert. 1985. Ocean carbon pumps: Analysis of relative strengths and efficiencies in ocean-driven atmospheric CO_2 changes. In: The Carbon Cycle and Atmospheric CO_2: Natural Variations, Archean to present, Geophysical Monograph 32, edited by E. T. Sundquist and W. S. Broecker. American Geophysical Union, Washington, D.C.:99-110.

Watson, Andrew. 1993. Air-sea gas exchange and carbon dioxide. In: The Global Carbon Cycle, edited by Martin Heimann. Springer-Verlag, Berlin:397-411.

Weiss, R. F. 1974. Carbon dioxide in water and seawater: the solubility of a non-ideal gas. Marine Chemistry 2:203-215.

Wiggert, Jerry, Tom Dickey, and Tim Granata. 1994. The effect of temporal undersampling on primary production estimates. Journal of Geophysical Research 99(C2): 3361-3371.

Williams, P. J. le B., K. R. Heinemann, J. Marra, and D. A. Purdie. 1983. Comparison of ^{14}C and O_2 measurements of phytoplankton production in oligotrophic waters. Nature 305:49-50.

Williams, P. J. leB., and J. E. Robertson. 1991. Overall planktonic oxygen and carbon dioxide metabolisms: the problem of reconciling observations and calculations of photosynthetic quotients. Journal of Plankton Research Supplement 13:S153-S169.

Williams, Peter J. leB. 1993. On the definition of plankton production terms. In: Measurement of Primary Production from the Molecular to the Global Scale: Proceedings of a Symposium held in La Rochelle, 21-24 April 1992, edited by William K. W. Li and Serge Y. Maestrini. ICES Marine Science Symposia 197:9-19.

Winn, Christopher D., Fred T. Mackenzie, Christopher J. Carrillo, Christopher L. Sabine, and David M. Karl. 1994. Air-sea carbon dioxide exchange in the North Pacific Subtropical Gyre: Implications for the global carbon budget. Global Biogeochemical Cycles 8(2):157-163.

World Meteorological Organization/World Data Centre for Greenhouse Gases. 1992. WMO WDCGG Data Report, Part A (Carbon Dioxide). World Meteorological Organization, Global Atmosphere Watch, World Data Centre for Greenhouse Gases. Japan Meteorological Agency. 776p.

Worthington, L. V. 1976. On the North Atlantic circulation. The Johns Hopkins Oceanographic Studies 6:1-110.

APPENDIX A
FORMULAS FOR THE SEASONAL MODEL

A.1 AIR-SEA GAS TRANSFER

We parameterize the CO_2 gas-exchange flux across the air-sea interface, F_{ex}, in mol m^{-2} s^{-1}, between the atmosphere and the mixed layer by means of the boundary-layer model of gas exchange (cf. Deacon 1977) given by equation (11) in the main text. The flux, F_{ex}, is a function of k_{ex}, the gas-exchange coefficient, in units of mol m^{-2} s^{-1} µatm^{-1}. It is also a function of pCO$_{2atm}$, the partial pressure of CO_2 in the atmosphere just above the surface, and of pCO$_{2oc}$, the partial pressure in the ocean immediately below. Both pressures are approximated by corresponding mixing ratios in parts per million by volume of dry air (ppm) (Heimann and Keeling 1989, 253-4).

The gas-exchange coefficient k_{ex} is evaluated at each time step as the product:

$$k_{ex}(t_{-1}) = P_v(t_{-1})\alpha_s(t_{-1})\rho_o, \tag{A.1}$$

where P_v denotes the piston velocity, in m s^{-1}; α_s, the solubility of CO_2 in seawater, in mol kg^{-1} µatm^{-1}; ρ_o, the density of seawater, in kg m^{-3}. The density, ρ_o, is set equal to a nominal constant value of 1026.2 kg m^{-3} (see Table 2), equal to the density of seawater with a salinity of 35 and a temperature of 20°C.

We compute α_s according to the expression of Weiss (1974)

$$
\ln(\alpha_s(t_{-1})10^6) = -60.2409 + \frac{9345.17}{T_{oc}(t_{-1})} + 23.3585 \ \ln \frac{T_{oc}(t_{-1})}{100}
$$
$$
+ S_o \left\{ 0.023517 - 0.023656 \ \frac{T_{oc}(t_{-1})}{100} + 0.0047036 \ \left[\frac{T_{oc}(t_{-1})}{100} \right]^2 \right\} \tag{A.2}
$$

where $T_{oc}(t_{-1})$ denotes the sea-surface temperature, here expressed in degrees Kelvin (degrees C + 273.15) at the beginning of the time step, and S_o, the salinity, is assumed to be constant (see Table 2).

We evaluate the piston velocity, P_v, as a function of wind speed according to the formula of Liss and Merlivat (1986) in a version parametrized by Heimann and Monfray (1989) thus:

$$
P_v(t_{-1}) =
$$
$$
\frac{\gamma}{360000} \cdot \left[0.17 \ (\alpha_p(T_{oc}))^{-\frac{2}{3}} \cdot \ U_{10}(t_{-1}) + 2.68 \ (U_{10}(t_{-1}) - 3.6) \ \cdot (\alpha_p(T_{oc}))^{-\frac{1}{2}} \right], \tag{A.3}
$$

where $U_{10}(t_{-1})$, in m s^{-1}, denotes the wind speed at 10 m above the sea surface. The first term of equation (A.3) reflects the dependency of the piston velocity on the wind speed for smooth surface conditions. The second term expresses the enhancement of the velocity caused by the presence of capillary waves at wind speeds greater than 3.6 m s^{-1} but less than

13.0 m s^{-1}. The expression is deemed valid for wind speed between 3.6 and 13.0 m s^{-1}. Thus it extends over the full range of averaged wind speeds listed in our Bermuda data set.

The term $\alpha_p(T_{oc})$ expresses the temperature dependency of P_v and was evaluated according to the formula of Heimann (personal communication) thus:

$$\alpha_p(T_{oc}) = \frac{Sc(T_{oc})}{600} = 10^{-6.706 + 1966 T_{oc}^{-1}}, \tag{A.4}$$

where the Schmidt number, Sc, is defined as the quotient of the kinematic viscosity of seawater and the coefficient of molecular diffusion of the gas, both quantities here taken to be functions of the temperature, $T_{oc}(t_{-1})$. The velocity, P_v, in equation (A.3) is adjusted to the global annual average radiocarbon uptake by the ocean by a constant factor, γ, set equal to 1.7447 as recommended by Keeling, Piper, and Heimann (1989) and Watson (1993, 410). With this adjustment, k_{ex} was computed to fall in the range 0.6 to $2.6 \cdot 10^{-9}$ mol $\text{m}^{-2} \text{ s}^{-1} \mu\text{atm}^{-1}$ compared to a global average of $2.1 \cdot 10^{-9}$ ($1.8106 \cdot 10^{-4}$ mol m^{-2} day^{-1} ppm^{-1}) quoted by Heimann and Keeling (1989, 254).

To calculate the change in the salinity-normalized concentration of DIC in the mixed layer that causes the gas-exchange flux, $\Delta sDIC_{ex}$, in mol kg^{-1}, we assume the CO_2 flux over each time step to mix homogenously throughout the layer, i.e.:

$$\Delta sDIC_{ex}(t) = \frac{F_{ex}(t) \cdot \Delta t}{MLD(t_{-1}) \cdot \rho_o}, \tag{A.5}$$

where Δt denotes the time step in seconds, $MLD(t)$ the mixed-layer depth, in m, and ρ_o, the density of seawater, having the same constant value as in equation (A.1).

The $^{13}CO_2$ air-sea transfer rate is computed independently of the corresponding flux of $^{13}CO_2 + {}^{12}CO_2$ by the formula given by equation (12) in the main text. In terms of $^{13}\delta$ ratios:

$$*F_{ex}(t_{-1}) = k_{ex}(t_{-1})R_s\alpha_{am} [pCO_{2atm}(t_{-1}) (^{13}\delta_{atm}(t_{-1}) + 1) -$$
$$\alpha_{eq}(t_{-1})pCO_{2oc}(t_{-1}) (^{13}\delta_{oc}(t_{-1}) + 1)], \tag{A.6}$$

where, as in the main text, $*F_{ex}(t)$ denotes the flux of ^{13}C in mol m^{-2} s^{-1}, R_s the $^{13}C/(^{12}C + {}^{13}C)$ ratio of the isotopic standard PDB and α_{am} the $^{13}C/^{12}C$ fractionation factor for CO_2 uptake by the surface ocean water, assumed to be constant. In addition, $^{13}\delta_{atm}$ denotes the reduced $^{13}C/^{12}C$ ratio of CO_2 in the atmosphere and $^{13}\delta_{oc}$ the corresponding ratio of sDIC. The equilibrium isotopic fractionation factor for gaseous CO_2 with respect to total DIC (i.e. DI^{13}C and DI^{12}C), denoted by α_{eq}, is assumed to vary with temperature.

The fractionation factor of gaseous CO_2 with respect of HCO_3^-, α_b, has been directly measured (Mook, Bommerson, and Staverman 1974, 175). Because total dissolved inorganic carbon consists of about 85% HCO_3^-, and the fractionation factors for the remaining inorganic carbon species are nearly the same as for HCO_3^-, we neglect the small influences of these on the isotopic equilibrium and equate α_{eq} with α_b, i.e.

$$\alpha_{eq}(t) \approx \alpha_b(t) = 1.02389 - \frac{9.483}{T_{oc}(t)} \tag{A.7}$$

where T_{oc} denotes the sea-surface temperature in degrees Kelvin as in equation (A.2). A more complete description of the derivation of these formulas is given in Heimann and Keeling (1989).

The $^{13}\delta$ value of F_{ex}, the gas-exchange flux, we denote by the reduced ratio, $^{13}\delta_{flux\,ex}$. It is evaluated (cf. equation (1) of the main text) by the expression:

$$^{13}\delta_{flux\,ex}(t) = \frac{*F_{ex}(t)}{(F_{ex}(t) - *F_{ex}(t)) \cdot r_s} - 1,$$

(A.8)

where $*F_{ex}/(F_{ex} - *F_{ex})$ expresses the $^{13}C/^{12}C$ ratio, r, of the flux.

The change in $^{13}\delta_{oc}$ in the mixed layer from the gas exchange, $\Delta^{13}\delta_{ex}$, is evaluated by assuming, as for DIC, that the flux of $^{13}C + ^{12}C$, summed over one time step and having the reduced ratio, $^{13}\delta_{flux\,ex}$, is mixed throughout the layer.

Using the additivity rule (equation (5) in the main text) the change in $^{13}\delta_{oc}$ in the mixed layer from the gas exchange is:

$$\Delta^{13}\delta_{ex}(t) = \frac{\Delta sDIC_{ex}(t)\,(^{13}\delta_{flux\,ex}(t) - {}^{13}\delta_{oc}(t_{-1}))}{sDIC(t_{-1}) + \Delta sDIC_{ex}(t)},$$

(A.9)

where $^{13}\delta_{oc}(t_{-1})$ denotes the observed $^{13}\delta$ of sDIC in the mixed layer at the beginning of the time step and $sDIC(t_{-1})$ the corresponding value of sDIC.

A.2 VERTICAL TURBULENT DIFFUSIVE TRANSPORT

The vertical turbulent diffusive flux of DIC, F_{diff}, in mol m^{-2} s^{-1}, is evaluated according to equation (13) of the main text as the product of the vertical diffusion coefficient, K_z, in m^2 s^{-1}, the vertical gradient in sDIC immediately below the mixed layer, $(dDIC/dz)_{lb}$, in mol kg^{-1} m^{-1}, and a nominal seawater density, ρ_o, in kg m^{-3}. Coefficient K_z, in turn, is determined as the product of a constant and the turbulent kinetic energy dissipation, ε, in m^2 s^{-3}, divided by the square of the buoyancy frequency, N, in s^{-1} according to equation (14). Following Oudot (1989) we have chosen ε to be equal to $2.0 \cdot 10^{-8}$ m^2 s^{-3}, a value that is representative for the upper thermocline at times of low wind speeds (Dillon and Caldwell 1980, 257). The local buoyancy frequency, N, is calculated from the vertical density gradient $(d\rho/dz)_{lb}$ at the base of the mixed layer. This gradient was computed separately for each station occupation by the formula:

$$N^2 = \frac{g}{\rho(t)} \left[\frac{d\rho(t)}{dz} \right]_{lb},$$

(A.10)

where g denotes the acceleration caused by gravity (9.81 m s^{-1}) and $\rho(t)$ the time varying density in the mixed layer, calculated from the observed temperature and salinity (Millero et al. 1980, 257).

The change in concentration of sDIC in the mixed layer over each time step due to vertical turbulent diffusion, $\Delta sDIC_{diff}$, in mol kg^{-1}, is given by a formula similar to equation (A.5):

$$\Delta sDIC_{diff}(t) = \frac{F_{diff}(t)\,\Delta t}{MLD(t_{-1})\,\rho_o}\,, \tag{A.11}$$

where $F_{diff}(t)$ denotes the diffusive flux in mol m^{-2} s^{-1}, Δt, the time step in seconds, and ρ_o is assigned the same value as in equation (A.1).

The flux of ^{13}C, $*F_{diff}$, into the mixed layer by vertical turbulent diffusion is computed by an expression analogous to equation (7), i.e.:

$$*F_{diff} = K_z(t_{-1})(d^{13}C/dz)_{lb}\,\rho_o, \tag{A.12}$$

where $(d^{13}C/dz)_{lb}$ denotes the vertical gradient in ^{13}C of DIC immediately below the mixed layer, and K_z is assumed to have the same value for ^{13}C as for ^{12}C + ^{13}C.

The $^{13}\delta$ value of F_{diff}, the diffusive flux, we denote by $^{13}\delta_{flux\ diff}$; it is defined in a manner similar to $^{13}\delta_{flux\ ex}$ (see equation (A.8)), i.e.:

$$^{13}\delta_{flux\ diff}(t) = \frac{*F_{diff}(t)}{(F_{diff}(t) - *F_{diff}(t)) - r_s} - 1. \tag{A.13}$$

The change in $^{13}\delta$ of DIC (caused by vertical turbulent diffusion, $\Delta^{13}\delta_{diff}$) is computed in a similar manner to $\Delta^{13}\delta_{ex}$ (see equation (A.9)), i.e.:

$$\Delta^{13}\delta_{diff}(t) = \frac{\Delta sDIC_{diff}(t) \cdot (^{13}\delta_{flux\ diff}(t) - {}^{13}\delta_{oc}(t_{-1}))}{sDIC(t_{-1}) + \Delta sDIC_{diff}(t)}. \tag{A.14}$$

Because we lack observations of the vertical gradient of the ^{13}C isotope as needed in equation (A.12), we have chosen to calculate the $^{13}\delta$ signature of the vertical diffusive flux, $^{13}\delta_{flux\ diff}$, directly from the observations of the vertical DIC and $^{13}\delta$ gradients as follows. Let $sDIC_1$ and $^{13}\delta_1$ denote the concentration of DIC and the reduced isotopic ratio, respectively at a short distance Δz_{lb} below the mixed layer, with corresponding values in the mixed layer, $sDIC_0$ and $^{13}\delta_0$. The gradients in sDIC and isotopic ratio in terms of the values just defined are:

$$(\delta sDIC/dz)_{lb} = (sDIC_1 - sDIC_0)/\Delta z_{lb} \tag{A.15}$$

and

$$(\delta^{13}\delta/dz)_{lb} = (^{13}\delta_1 - {}^{13}\delta_0)/\Delta z_{lb}. \tag{A.16}$$

The difference $^{13}\delta_1 - {}^{13}\delta_0$ is related to the corresponding concentration difference, $sDIC_1 - sDIC_0$, assuming that sDIC with an isotopic ratio $^{13}\delta_{flux\ diff}$ has been added to a pool of inorganic carbon of concentration $sDIC_0$, to increase its concentration to $sDIC_1$. Calculating the isotopic relationships for this increase by the additivity rule (equation 5), we obtain:

$$^{13}\delta_0 sDIC_0 + {}^{13}\delta_{flux\ diff}(sDIC_1 - sDIC_0) = {}^{13}\delta_1 sDIC_1, \tag{A.17}$$

whence

$$^{13}\delta_{flux\ diff} = {}^{13}\delta_1 + \frac{(d^{13}\delta/dz)_{lb}}{(dsDIC/dz)_{lb}} \cdot sDIC_0, \tag{A.18}$$

where we solved equation (A.17) for $^{13}\delta_{flux\ diff}$ with substitutions from equations (A.15) and (A.16) to eliminate $^{13}\delta_0$ and $sDIC_1$ from the resulting expression.

In the limit that Δz approaches zero, $^{13}\delta_1$ approaches $^{13}\delta_0$, and

$$^{13}\delta_{flux\ diff} = {}^{13}\delta_0 + \frac{(d^{13}\delta/dz)_{lb}}{(dsDIC/dz)_{lb}} \cdot sDIC_0. \tag{A.19}$$

Thus, $^{13}\delta_{flux\ diff}$ can be determined knowing only the concentration and isotopic ratio of sDIC in the mixed layer and their gradients immediately below. In the model computations, equation (A.19) has been used directly to calculate $^{13}\delta_{flux\ diff}$ instead of equations (A.16) and (A.14). Because both the DIC and $^{13}\delta$ gradient are assumed to be constant (0.45 µM kg^{-1} m^{-1} and -0.0021 ‰ m^{-1}, respectively), and $^{13}\delta_0$ and sDIC$_0$ vary only slightly, the resulting isotopic signature of vertical diffusion, $^{13}\delta_{flux\ diff}$, is nearly constant (-7.9 ‰).

A.3 VERTICAL ENTRAINMENT

Entrainment of water into the mixed layer from the submixed layer below occurs when the mixed layer deepens episodically as a result of wind-induced turbulent stirring of the upper ocean, or when cooling at the sea surface reduces the static stability of the water column. In general, the associated flux of DIC into the mixed layer, F_{ent}, is expressed as a product of the instantaneous difference in the concentration of DIC across the base of that mixed layer and the rate of increase, dMLD/dt, of that layer, which must be positive for entrainment to occur (Phillips 1977). Because we lack data at Station S to estimate the magnitude of the instantaneous difference in DIC, we have assumed that this quantity can be approximated from the observed gradient in DIC in the thermocline below the mixed layer. Thus, we replace the presumed discontinuity at the base of the mixed layer by an assumed continuous profile below the mixed layer, the latter having a gradient $(dDIC/dz)_{lb}$, which is constant in time and depth during the period under consideration. We further assume that entrainment occurs as a series of episodes during which the mixed layer abruptly increases in depth. It follows that the average difference in DIC during one episode of entrainment is:

$$\overline{\Delta DIC_{base}} = \frac{1}{2} (dDIC/dz)_{lb}\Delta MLD_{ent}, \tag{A.20}$$

where ΔMLD_{ent} denotes the increase in depth. If the next episode of entrainment takes place after a recurrence interval, Δt_{ent}, the flux of DIC averaged over this interval, $\overline{F_{ent}}$, is

$$\overline{F_{ent}} = \frac{1}{2} (dDIC/dz)_{lb} \frac{\Delta MLD_{ent}^2}{\Delta(t)_{ent}} . \tag{A.21}$$

Because the mixed-layer depth is expressed in the model by a slowly varying, two-harmonic function, the increase in depth, ΔMLD_{ent}, is roughly proportional to Δt_{ent} over the relatively short time intervals that comprise a few successive episodes of entrainment. Therefore, the flux, F_{ent}, by equation (A.21), tends to increase in direct proportion to Δt_{ent} and cannot be set arbitrarily, for example, as equal to the model time step, Δt.

The prorated effect of entrainment for a single time step, say from t_{-1} and t, is assumed to involve a period of entrainment beginning at a time:

$$t_{pre} = t - \Delta t_{ent}/2 - \Delta t/2,\tag{A.22}$$

and ending at

$$t_{post} = t + \Delta t_{end}/2 - \Delta t/2,\tag{A.23}$$

where the episode is centered, as is the time step, at time $(t - \Delta t/2)$. For each additional proration step, the time of entrainment is assumed to shift forward by one time step, Δt. The change in depth of the mixed layer during an episode of entrainment is computed by the expression:

$$\Delta MLD_{ent}(t) = MLD(t_{post}) - MLD(t_{pre}).\tag{A.24}$$

The average concentration of sDIC in the water parcel incorporated into the mixed layer during an episode, $sDIC_{lb}(t)$, is computed from $(dDIC/dz)_{lb}$, by the expression:

$$sDIC_{lb}(t) = sDIC_{lb}(t_{pre}) + \frac{1}{2} \cdot (dDIC/dz)_{lb} \cdot \Delta MLD_{ent}(t),\tag{A.25}$$

where $sDIC(t_{pre})$ denotes the sDIC concentration in the mixed layer at the beginning of the entrainment episode.

The sDIC of the mixed layer after an episode of entrainment, $sDIC(t_{post})$, is computed assuming that entrained water mixes completely with water of the mixed layer, neglecting changes in sDIC by any other process, i.e:

$$sDIC(t_{post}) = \frac{sDIC(t_{pre})\,MLD(t_{pre}) + sDIC_{lb}(t)\,\Delta MLD_{ent}(t)}{MLD(t_{pre}) + \Delta MLD_{ent}(t)}.\tag{A.26}$$

The prorated change in sDIC entrainment during a single time step is therefore given by

$$\Delta sDIC_{ent}(t) = \frac{(sDIC(t_{post}) - sDIC(t_{pre}))\,\Delta t}{\Delta t_{ent}}.\tag{A.27}$$

The $^{13}\delta$ of DIC in the water parcel incorporated during the time step, $^{13}\delta_{lb}$ (cf. equation A.25), is evaluated by:

$$^{13}\delta_{lb}(t) = {}^{13}\delta_{lb}(t_{pre}) + \frac{1}{2} \cdot (d^{13}\delta/dz)_{lb} \cdot \Delta MLD_{ent}(t),\tag{A.28}$$

where $^{13}\delta(t_{pre})$ denotes the reduced isotopic ratio of sDIC in the mixed layer at the beginning of the entrainment episode and, as in the case for DIC, $(d^{13}\delta/dz)_{lb}$, the vertical gradient in $^{13}\delta$ of DIC in the layer of water immediately below the mixed layer, is assumed to be constant in time. The $^{13}\delta$ of DIC in the mixed layer after an episode of entrainment, $^{13}\delta(t_{post})$, is computed assuming, as in equation (A.26), that entrained water mixes completely with water of the mixed layer neglecting changes in $^{13}\delta$ by other processes. Hence,

$$^{13}\delta(t_{post}) = \frac{^{13}\delta(t_{pre})PROD(t_{pre}) + {}^{13}\delta_{lb}(t)\, PROD(t)}{PROD(t_{pre}) + PROD(t)}, \tag{A.29}$$

where

$$PROD(t_{pre}) = sDIC(t_{pre})MLD(t_{pre}), \tag{A.30}$$

and

$$PROD(t) = sDIC_{lb}(t)\Delta MLD_{ent}(t). \tag{A.31}$$

The prorated change in $^{13}\delta_{oc}$ by entrainment during a single time step (cf. equation (A.27)) is given by:

$$\Delta^{13}\delta_{ent}(t) = \frac{(^{13}\delta(t_{post}) - {}^{13}\delta(t_{pre}))\,\Delta t}{\Delta t_{ent}}. \tag{A.32}$$

The entrainment flux, F_{ent}, is computed by equation (18) of Subsection 6.6 of the main text, using the value of $\Delta sDIC_{ent}$ given by equation (A.27).

APPENDIX B
THREE-DIMENSIONAL GLOBAL OCEAN TRACER
TRANSPORT MODEL OF BACASTOW AND
MAIER-REIMER (1991)

We made use of the results of a three-dimensional ocean carbon model for estimating vertical gradients of DIC and its isotopic ratio below the mixed layer ($[dDIC/dz]_{lb}$ and $[d^{13}\delta/dz]_{lb}$). We also used this model to compare the flux estimates of our seasonal box model with those of a higher order model.

This three-dimensional model employs the circulation field of a 10-layer seasonally averaged global circulation model obtained from the Max Planck Institute of Meteorology, Hamburg, Germany, with a resolution of 5° x 5°, i.e. about 550 km near the equator. The carbon cycle model is as described by Bacastow and Maier-Reimer [1990], except that vertical convection is parameterized by diffusion, and a cycle of dissolved organic carbon is included, as described by Bacastow and Maier-Reimer [1991]. The version of the model used is nonseasonal with a surface layer 115 m deep. For our computations we used the results for the 475 by 550 km grid box encompassing Bermuda.

From a pair of runs of the model from the year 1740 to 1985 (Bacastow, private communication), the component gradients of the marine carbon cycle were estimated from the model predictions of the three-dimensional fields of DIC and $^{13}\delta$. Two model runs, A and C, were first spun up to 1740. These differ only in that A refers to an ocean with no biota (cf. Broecker and Peng 1982) while C retains a biota. Both were set to be in equilibrium in 1740 with a well-mixed atmosphere of 280 ppm CO_2. Afterward, the model runs were continued with the addition of fossil fuel CO_2 to the model atmosphere, according to estimates based on historical data by Keeling et al. (1989). A biota source resulting from land clearing was adjusted so that the modeled increase of atmospheric CO_2 matched the observations from ice core data and direct measurements by a procedure called deconvolution (Keeling et al. 1989, 195 and Appendix D, therein). The runs, extended to 1985, take model run A to B, and run C to D.

Gradients of $^{13}\delta$ and DIC in model-run A result exclusively from temperature-dependent solubility of CO_2 (the solubility pump) and from temperature-dependent fractionation. The changes in the gradients caused by the buildup of fossil fuel (the Suess effect) were estimated for both runs B and D by calculating the difference between the results for 1740 and 1985. The effect of the ocean biota was similarly calculated by the difference between model runs A and C for 1740 or between runs B and D for 1985. The differences between the estimates of the effect of the ocean biota in the absence or presence of the Suess effect and, similarly, the differences between the estimates of the Suess effect in the absence or presence of the ocean biota, are both small. This confirms that the three processes influence the fields of DIC and $^{13}\delta$ in the three-dimensional model nearly independently. These nonseasonal predictions cannot, however, be directly compared to the actual gradients below the mixed layer at Station S because of its variable depth. We weighted the computed overall gradient between the first two levels (centered at 56 and 156 m depth) by 47.6% and the gradient between the second and third level (the latter centered at 275 m) by 52.4% in order to match the observed overall gradient of 0.45 μM kg^{-1} m^{-1}. The same prorating factor was then applied to determine the corresponding isotopic gradients. The computed overall gradients and the gradients of the individual components are shown in Table 3.

APPENDIX C
SENSITIVITY TESTS

C.1 INTRODUCTION

Here we present details of the results of sensitivity tests carried out using our model of the seasonal cycle of carbon in the waters near Bermuda. These tests are summarized in Subsection 8.1 of the main text. They are with respect to a standard case defined by the data presented in Tables 1, 2, 4, and 5 and discussed in Section 7.

C.2 AIR-SEA GAS TRANSFER

To test the sensitivity of our calculations to errors in the gas-exchange coefficient, k_{ex}, we varied the factor, γ, which adjusts by a constant the relationship between wind speed and piston velocity, originally specified by Liss and Merlivat (1986). When γ varies between 1.0 and 2.0 the net uptake of atmospheric CO_2 by the mixed layer, F_{ex}, integrated over the annual cycle varies, respectively, between 12.3 and 24.6 gC m^{-2} and the net biological exchange flux of carbon, F_{bio}, over the annual cycle varies by almost the same amount, from -1.8 to -13.7 gC m^{-2} (shown in Table 6). Thus the model predicts that whatever error exists in computing the air-sea flux is almost directly passed on as an error in net community production. With respect to the standard case (γ equal to 1.74), the lack of closure of $sDIC_{calc}$ over the annual cycle (see Subsection 5.2, above) is slightly worse than the standard case using the factor 2.0, still worse using the factor 1.0 (see Figure 14).

As noted in Section 4, above, intercalibrations with direct measurements of pCO_{2oc} indicate that our data, computed from temperature, salinity, dissolved inorganic carbon, and alkalinity measurements using thermodynamic relationships, if not corrected, would be about 10 ppm too high (Lueker et al. submitted manuscript). As a sensitivy test we increased pCO_{2oc} in the mixed layer uniformly over the whole seasonal cycle by 10 ppm to restore the originally calculated values. This adjustment decreases annual F_{ex} from 21.4 gC m^{-2} in the standard case to 14.9 gC m^{-2}, while annual F_{bio} decreases from -10.7 gC m^{-2} to -7.6 gC m^{-2} (Table 6). Annual F_{bio} decreases in absolute magnitude by only about half of the decrease in F_{ex} because of an associated isotopic adjustment. The lack of closure of $sDIC_{calc}$ over the annual cycle in the standard case (-1.8 µmol kg^{-1}) is made considerably worse (-7.4 µmol kg^{-1}), because the reduced uptake of DIC in the water by gas exchange is only partially compensated for by decreased removal of carbon by biological exchange.

To test the sensitivity of the model to the degree of seasonality in air-sea exchange, we kept the gas-exchange coefficient, k_{ex}, constant over the annual cycle. We chose a value (see Table 6) such that annual F_{ex} remained the same as for the standard case, and thus slightly higher than is consistent with the global mean gas-exchange rate of Heimann and Keeling (1989) of $2.0956 \cdot 10^{-9}$ mol m^{-2} s^{-1} ppm^{-1}, based on observations. Annual F_{bio} increases from -10.7 gC m^{-2} to -15.9 gC m^{-2} (Table 6). The difference in extreme values of F_{ex} over the annual cycle increases from 0.164 gC m^{-2} day^{-1} to 0.200 gC m^{-2} day^{-1}, as a result of a greater release of CO_2 to the air during summer. As shown in Figure 14, this sensitivity test results in a very poor closure of $sDIC_{calc.}$

We also investigated the importance of seasonal variations in temperature of the surface mixed layer, T_{oc}, as it applies to the equations governing air-sea exchange, including those involving isotopic fractionation. We retained the representation of pCO_{2oc} given by the harmonic function prescribed in Table 1, however. Thus, the sensitivity of pCO_{2oc} to temperature variations is not included in this test. If T_{oc} for the whole year is set equal to the average observed temperature (23.03°C) in equations (A.2), (A.3), (A.4), and (A.7) of Appendix A.1, annual F_{ex} increases only slightly from 21.4 gC m^{-2} to 21.5 gC m^{-2}. Also, annual F_{bio} is only slightly changed.

An additional sensitivity test was carried out to clarify the importance of isotopic fractionation during air-sea exchange. The most critical of these factors is the extent of isotopic disequilibrium at the air-sea boundary brought about by the oceanic absorption of CO_2 from fossil fuel combustion (Suess effect). If the temperature, T_{oc}, in equation (A.7) is set equal to the average observed value of 23.03°C, the equilibrium fraction factor, α_{eq}, is found on average to be 0.99187. At isotopic equilibrium,

$$\alpha_{eq} = r_{atm} / r_{oc}. \tag{C.1}$$

Hence, in view of equation (1);

$$\alpha_{eq} = (^{13}\delta_{atm} + 1)/(^{13}\delta_{oc} + 1). \tag{C.2}$$

Thus, in the absence of the Suess effect, and disregarding seasonal variations, the reduced isotopic ratio of atmospheric CO_2 at Station S, $^{13}\delta_{atm}$, would tend toward the equilibrium value -6.615‰ rather than -7.700‰, as determined from observations and assuming that $^{13}\delta_{oc}$ has the value 1.527‰ (see Table 1). If we neglect the Suess effect in the model by making $^{13}\delta_{atm}$ less negative by 1.085‰, annual F_{bio} is drastically reduced from -10.7 gC m^{-2} to 0.7 gC m^{-2}. Because there are no changes from the standard case for the fluxes of the other three processes, the closure of $sDIC_{calc}$ over the annual cycle (16.5 μmol kg^{-1}, see Figure 14) is very poor.

Neglect of the Suess effect has a further curious consequence. Two small cusps occur in the curves of $\Delta sDIC_{bio}/\Delta t$ and F_{bio} (Figures 9, 10, 12, and 13) on days 172 and 272 in the standard case. (The first cusp is, however, too small to see in the plots.) On these two days the CO_2 partial pressure difference at the air-sea boundary momentarily becomes zero. The two cusps reflect that even though this is the case, the pressure differences of the individual species $^{13}CO_2$ and $^{12}CO_2$ are not zero, as they would be if an isotopic equilibrium prevailed. The actual pressure differences are almost surely too erratic for such a transient isotopic disequilibrium ever to be apparent in observations, but because input data are smoothed in our model calculations, small abrupt changes in the time rate of change in $\Delta sDIC_{bio}/\Delta t$ and F_{bio} arise from the opposition of these two isotopic air-sea fluxes. When the Suess effect is removed from the calculations, these cusps disappear.

C.3 VERTICAL TRANSPORT

The importance of a correct estimation of the upward flux of DIC into the mixed layer by diffusion is seen by performing a calculation in which $\Delta sDIC_{diff}$ is set equal to zero. The net biological exchange flux, F_{bio}, becomes less negative (or more positive) throughout the year (see Table 6) with a net value over the annual cycle of -4.2 gC m^{-2} instead of -10.7 gC m^{-2}. Doubling the diffusive flux throughout the year has an equal but opposite effect (annual F_{bio} of -17.2 gC m^{-2}). The very poor closure of the annual cycle of $sDIC_{calc}$, in opposite directions for those two extreme cases, suggests that our chosen rate is nearly correct.

In the standard case the recurrence interval of entrainment was set to be 8 days on the basis of only very limited direct information. Changing the entrainment recurrence interval, Δt_{ent}, has an almost proportional effect on the entrainment flux, F_{ent}. Because the annual effect of this flux is small, however (3.2 gC m^{-2}), changing the interval of entrainment during deepening to 4 or 12 days has only a minor influence on the model predictions.

To test the hypothesis that the predictions of our model are sensitive to the magnitude of vertical transport, irrespective of whether this occurs by vertical diffusion or entrainment, we assumed a low diffusion coefficient and adjusted the entrainment recurrence interval, Δt_{ent}, in a manner that the combined vertical transport was of the same magnitude as in the standard case. We chose a value of $0.1 \cdot 10^{-4}$ m^2 s^{-1} for the vertical diffusion coefficient, K_z, in accordance with a recent estimate by Ledwell, Watson, and Law (1993) and adjusted Δt_{ent} to 60 days. In this test, the annual diffusive flux, F_{diff}, decreases to 1.7 gC m^{-2}, whereas the annual entrainment flux, F_{ent}, increases to 15.8 gC m^{-2}, summing to a combined annual vertical transport of 17.5 gC m^{-2} in comparison to 18.3 gC m^{-2} in the standard case. Annual F_{bio} changes very little as a result of opposing changes during the shoaling period and the deepening period. The lack of closure of $sDIC_{calc}$ over the annual cycle is almost the same as for the standard case and the shape of the annual curve of sDIC is similar to that of the standard case (cf. Figure 14).

Because our estimate of the observed vertical gradient of $^{13}\delta$ below the mixed layer, $(d^{13}\delta/dz)_{lb}$, is very uncertain, we employed in our standard calculation an estimate derived from the three-dimensional ocean transport model of Bacastow and Maier-Reimer (1991), as described in Appendix B. The model prediction is -0.0021‰ m^{-1}, whereas the average of four vertical profiles indicated a gradient of approximately -0.003‰ m^{-1} (see Section 6.3). Substituting this latter value for the model estimate increases annual F_{bio} from -10.7 gC m^{-2} to -14.1 gC m^{-2}, with nearly equal contributions to the differences from the standard case for shoaling and deepening. The lack of closure of $sDIC_{calc}$ over the annual cycle worsens from -1.8 μM kg^{-1} for the standard case to -5.4 μM kg^{-1}.

Our determination of the mixed-layer depth assumes a constant difference, $\Delta\sigma_t$, between the density of seawater at the sea surface and the density, which demarks the beginning of the pycnocline below, according to the criterion of Levitus (1982). As noted earlier, this criterion leads to estimates of depth in good agreement with subjective estimates for most of the annual cycle. Sometimes, however, during the shoaling period, no distinct pycnocline could be seen. The estimate of the depth is then arbitrary. In sharp contrast to the gradual deepening process, shoaling appears to have been highly erratic, with the mixed layer shifting up and down rapidly, as seen in the time series that we proposed as the basis for Figure 2.

The use of a harmonic fit tends to smooth out this variability and, by describing a gradual shoaling, prescribes the mixed layer to be deeper during May and June than is indicated from the individual density profiles. As a sensitivity test we fitted the seasonally composite data for mixed-layer depth with a polynomial hermite spline, forced to describe a rapid shoaling in April, as shown in Figure 15. The rates of change in concentration, $\Delta sDIC_i / \Delta t$, previously shown in Figure 9 for the standard case, are shown based on the Hermite fit in Figure 16. The assumption of a shallower mixed layer in May leads to large increases in all of the calculated fluxes, except for entrainment, modeled to be zero during shoaling.

As a second sensitivity test we challenged our harmonic fit to the individual estimates by increasing the mixed-layer depth by 30 meters throughout the year. As seen in Figure 2, the original fit to the data falls short of reproducing the envelope of extreme mixed-layer depths in autumn and winter. The relatively short periods in which the mixed layer is temporarily deeper than the trend curve determined by the harmonic fit, may cause most of the vertical transport because of the greater turbulence and vertical mixing required to produce extreme depths. By increasing the mixed-layer depth in the model by 30 meters throughout the year the adjusted harmonic fit comes considerably closer to matching these deeper events, although it considerably exaggerates the depth in the summer. Annual F_{bio} decreases only slightly from the standard case. The lack of closure of $sDIC_{calc}$ over the annual cycle was worsened slightly from -1.8 µM/kg to -4.1 µM/kg.

C.4 BIOLOGICAL EXCHANGE

We estimated the uncertainty in α_{org} from the observed variability of the $^{13}\delta$ of marine phytoplankton. Over the temperature range found at Bermuda (18 - 28° C), Rau, Takahashi, and Des Marais (1989) reported a variation of $^{13}\delta_{org}$ of ± 2‰. We included the possibility of an overlooked small fractionation of ±1‰ during respiration, thus resulting in an overall uncertainty in α_{org} not likely to exceed ± 3‰. A fractionation greater by 3‰ (α_{org} - 1 changed from -22.0‰ to -25.0‰) negligibly decreases the net biological exchange flux, F_{bio}, over the annual cycle from the standard case with only slightly greater differences for times of shoaling and deepening (see Table 6). As expected, a lesser fractionation by 3‰ has essentially an opposite effect. No other calculated fluxes are affected by this isotopic parameter. Evidently, uncertainty in α_{org} introduces only small errors in the model calculation of net community production.

TABLE 1

Harmonic fitting coefficients of equation (7)

Parameter*	S_0	a_1	b_1	a_2	b_2	a_3	b_3	R^2
Temperature (°C)	23.034	-3.526	-2.088	0.546	0.262	–	–	0.978
sDIC (μM kg^{-1})	2029.86	14.52	0.48	1.76	1.11	–	–	0.941
$^{13}\delta_{oc}$ (‰)	1.527	-0.096	-0.002	-0.026	0.004	–	–	0.621
$^{13}\delta_{atm}$ (‰)	-7.700	-0.2385	-0.0619	0.1013	-0.035	–	–	–
pCO_{2oc} (ppm)†	324.86	-22.27	-27.13	8.31	5.17	–	–	0.958
pCO_{2atm} (ppm)	348	3.304	0.817	-1.365	0.494	–	–	–
Mixed layer depth (m)	70.840	47.298	53.872	18.759	-4.243	-1.631	-2.533	0.716
K_z (10^{-4} m^2 s^{-1})	0.868	0.949	0.379	0.133	-0.397	-0.085	-0.030	0.674
Wind speed (m s^{-1})	7.173	0.764	1.894	0.057	-0.279	–	–	0.984

* For definitions of symbols see Table 2.

† pCO_{2oc} reduced by 10 ppm from original calculation (see Section 4).

Table 2

Input data to the seasonal model

Parameter	Value or Range	Source of Data
Temperature of mixed layer (T_{oc})	18 - 29°C	Observed, two harmonic fit (see Table 1)
Average Salinity (S_o)	36.452	Observed, average value of time series
DIC of mixed layer normalized to S_o (sDIC)	2000 to 2050 μmol kg^{-1}	Observed, two harmonic fit (see Table 1)
$^{13}\delta$ of DIC of mixed layer ($^{13}\delta_{oc}$)	1.3 to 1.8‰	Observed, two harmonic fit (see Table 1)
$^{13}\delta$ of atmospheric CO_2 ($^{13}\delta_{atm}$)	-7.25 to -7.95‰	Observed at La Jolla for 1987, two harmonic fit (see Table 1)
$^{13}\delta$ of phytoplankton in mixed layer ($^{13}\delta_{org}$)	-20.7 to -20.3‰	Rau et al. [1989] (see equation (17a))
pCO_2 at the sea surface and in the mixed layer (pCO_{2oc})	290 to 390 ppm	Observed, two harmonic fit (see Table 1)
pCO_2 of atmosphere ($pCO_{2\,atm}$)	340 to 355 ppm	Observed at La Jolla for 1987, two harmonic fit (see Table 1)
Mixed-layer depth (MLD)	15 to 270 m	Observed, three harmonic fit (see Table 1)
Gas-exchange coefficient (k_{ex})	0.6×10^{-9} to 2.6×10^{-9} mol m^{-2} s^{-1} μatm^{-1}	Calculated from wind speed, gas solubility and density of sea water (see equation (A.1))
Vertical diffusion coefficient (K_z)	5×10^{-6} to 5×10^{-4} m^2 s^{-1}	Calculated from vertical density gradient immediately below the mixed layer, three harmonic fit
Average density of seawater in mixed layer	1026.2 kg m^{-3}	From observations
Photosynthetic fractionation factor (α_{org})	0.97805 to 0.97745	Rau et al. (1989) (varies with [$CO_2]_{aq}$, see equation (17a))
Atmospheric CO_2/oceanic DIC equil. fractionation factor (α_{eq})	0.99144 to 0.99236	Mook et al. (1974, 175)

Table 2 (continued)

Parameter	Value or Range	Source of Data
Air-sea kinetic fractionation factor (α_{am})	0.99820	Keeling et al. (1989, 189)
$^{13}C/(^{13}C + {}^{12}C)$ of standard (R_s)	0.011112328	Heimann and Keeling (1989, 261)
$^{13}C/^{12}C$ of standard (r_s)	0.0112372	Mook and Grootes (1973, 296)
Vertical DIC gradient† ($dDIC/dz)_{lb}$	0.45 μmol kg^{-1} m^{-1}	Estimated from vertical profiles of DIC
Vertical $^{13}\delta$ gradient† ($d^{13}\delta/dz)_{lb}$	-0.0021‰ m^{-1}	Estimated from three-dimensional model of Bacastow and Maier-Reimer (1991)
Wind speed at 10 m (U_{10})	4 to 9 m s^{-1}	Musgrave et al. (1988) Monthly average wind speed two harmonic fit
Piston velocity adjustment factor (g)	1.7447	Keeling, Piper, and Heimann (1989, 316)

† z positive downward.

65

TABLE 3

Vertical DIC and $^{13}\delta$ gradients below the mixed layer as calculated by the three-dimensional ocean-tracer transport model by Bacastow and Maier-Reimer (1991)

(see Appendix B for details)

Processes	$(dDIC/dz)_{lb}$ [μmol kg^{-1} m^{-1}]	$(d^{13}\delta/dz)_{lb}$ [‰ m^{-1}]
Temperature dependent fractionation and CO_2 solubility	0.16	0.0004
Buildup of industrial CO_2 (Suess effect)	-0.07	0.0014
Remineralization of organic matter	0.36	-0.0039
Overall gradients	0.45	-0.0021

TABLE 4

Local changes in concentration of the inorganic carbon system in the mixed layer at Station S*

Process	Time Period		
	Annual ($gC\ m^{-3}$)	Shoaling† ($gC\ m^{-3}$)	Deepening ($gC\ m^{-3}$)
Inferred from model:			
Biological exchange, $\Delta sDIC_{bio}$	-0.418	-0.294	-0.124
Air-sea gas exchange, $\Delta sDIC_{ex}$	0.156	0.098	0.058
Vertical diffusive transport, $\Delta sDIC_{diff}$	0.199	0.120	0.079
Vertical entrainment, $\Delta sDIC_{ent}$	0.040	0.000	0.040
Calculated sum of components	-0.022	-0.075	0.053
Observed from the seasonal cycle of sDIC:			
Sum of components	0.000	-0.169	0.169

* Positive values indicate fluxes into the mixed layer.
† From 19 February to and including 9 July. (All other times are deepening.)

TABLE 5

Vertically integrated rates of change of the inorganic carbon system in the mixed layer at Station S

Process	Time Period		
	Annual (gC m^{-2})	Shoaling (gC m^{-2})	Deepening (gC m^{-2})
Inferred from model:			
Biological exchange, F_{bio}	-10.68	-19.02	8.34
Air-sea gas exchange, F_{ex}	21.45	10.08	11.37
Vertical diffusive transport, F_{diff}	15.18	9.49	5.69
Vertical entrainment, F_{ent}	3.15	0.00	3.15
Calculated sum of components, F_{calc}	29.10	0.54	28.55
Observed from the seasonal cycle of sDIC:			
F_{obs} *	26.94	-6.45	33.39
F_{tot} †	0.0	-3711.68	3711.68
$F_{tot} - F_{obs}$	-26.94	-3705.23	3678.29

* Change with respect to concentration of DIC only (see equation (20)).
† Change also with respect to mixed-layer depth (see equation (21)).

TABLE 6

Computed net community production in mixed layer and lack of closure of sDIC at Station S

Sensitivity Test*	Net Community Production			Annual
	Annual (gC m^{-2})	Shoaling (gC m^{-2})	Deepening (gC m^{-2})	$\Delta sDIC_{calc}$ (μM kg^{-1})
Standard Case	-10.7	-19.0	8.3	-1.8
Air-sea gas transfer				
Piston velocity adj. factor, $\gamma = 1$	-1.8	-15.5	13.7	3.8
Piston velocity adj. factor, $\gamma = 2$	-13.7	-20.2	6.5	-3.8
pCO_{2oc} increased by 10 ppm	-7.6	-17.7	10.1	-7.4
$k_{ex} = 2.4283 \times 10^{-9}$ mol m^{-2} s^{-1} ppm^{-1} (constant value)	-15.9	-21.9	6.0	-26.6
$T_{oc} = 23°C$ (constant value)	-11.6	-20.5	9.0	1.0
$^{13}\delta_{atm}$ less negative by 1.085‰ (neglect of the Suess effect)	0.7	-14.8	15.5	16.5
Vertical transport and mixed-layer depth				
Diffusion doubled	-17.2	-23.0	5.9	7.4
$\Delta sDIC_{diff} = 0$ (no diffusion)	-4.2	-15.0	10.8	-11.1
$\Delta t_{ent} = 4$ days	-10.0	-19.0	9.0	-2.7
$\Delta t_{ent} = 12$ days	-11.3	-19.0	7.7	-1.0
$K_z = 0.1 \cdot 10^{-4}$ m^2 s^{-1} and $\Delta t_{ent} = 60$ days	-10.4	-15.3	4.8	-1.4
$(d^{13}\delta/dz)_{lb} = -0.0030‰$ m^{-1} (observed gradient)	-14.1	-20.8	6.7	-5.4
MLD determined by Hermite function	-12.2	-15.3	3.2	2.5
MLD increased all years by 30 m	-11.1	-22.8	11.7	4.1
Biological exchange				
α_{org} more negative by 3‰	-11.2	-17.7	6.5	0.4
α_{org} less negative by 3‰	-10.0	-20.8	10.8	-4.7

* For definitions of symbols see Table 2

TABLE 7

Estimates of annual biological production in the Sargasso Sea

Author	Type of Production	Estimate (gC m^{-2} yr^{-1})	Location	Method
Menzel and Ryther (1960)	Primary Production	72	Station S	^{14}C incubation
Platt and Harrison (1985)	New Production	25α	Station S	nitrate based f-ratio
Jenkins and Goldman (1985)	New Production	35 - 50β	Station S	oxygen cycle
Musgrave et al. (1988)	New Production	25 - 34β	Station S	oxygen cycle
Spitzer and Jenkins (1989)	New Production	37 - 48β	Station S	oxygen, argon, and helium cycles
Altabet (1989a)	Downward flux of sinking PN	18γ	BATS Station	sediment traps
	Downward flux of suspended PN	12γ	BATS Station	PN cycle
Lohrenz et al. (1992)	Primary Production	110 - 144δ	BATS Station	^{14}C incubation
	Downward flux of POC	14ε	BATS Station	sediment traps
Carlson et al. (1994)	Downward flux of DOC	13ζ	BATS Station	DOC cycle
Fasham et al. (1990)	New Production in the mixed layer	37γ	Station S	mixed-layer ecosystem model
Fasham et al. (1993)	New Production	35 - 52γ	Station S	three-dimensional ecosystem model
This study, 1995	Net Community Production in the mixed layer	11	Station S	concentration and ^{13}C/^{12}C ratio of DIC
	Net Community Production in the euphotic layer	18	Station S	extrapolation of mixed-layer value

α Based on weighted average f-ratio of 0.31 and primary production of Menzel and Ryther (1960).

β Conversion of oxygen into carbon based on average PQ = 1.41 proposed by Takahashi et al. (1985).

γ Conversion of nitrogen to carbon based on C:N ratio of (122:16) of Takahashi et al. (1985).

δ Annual primary production for the years 1989 and 1990, respectively.

ε Estimated fluxes at 100 m based on the reported values at 150 m and assumption that fluxes at that depth are approximately 67 % of the fluxes at 100 m according to relationship by Martin et al. (1987).

ζ Average value for years 1992 and 1993, respectively.

TABLE A.1

Summary of Station S carbon data as used in this study

Date [YYMMDD]	Depth [m]	Sal	Temp [°C]	DIC [μmol kg⁻¹]	Alk [μmol kg⁻¹]	pCO₂ [ppm]	Orig ¹³δ [per mil]	Actual ¹³δ [per mil]
830912	1	36.150	28.01	1999.54	–	373.27	–	–
830912	10	36.146	27.24	2000.52	–	364.89	1.80	1.54
831010	1	35.950	26.35	1987.33	2346.44	347.90	1.62	1.62
831010	10	35.945	26.35	1986.95	2346.72	346.93	–	1.60
840127	1	36.382	20.20	2031.94	2370.23	312.17	–	1.50
840127	10	36.381	20.41	2030.00	–	306.09	–	1.35
840309	1	36.463	19.81	2043.46	–	312.15	–	–
840309	10	36.466	19.88	2042.87	–	311.99	–	–
840403	1	36.484	19.61	2046.56	–	312.66	–	1.43
840403	10	36.468	19.49	2044.80	–	309.74	–	1.39
840430	1	36.479	19.74	2043.68	–	310.55	1.46	1.46
840430	10	36.481	20.05	2043.89	–	314.56	1.46	1.46
840628	10	36.300	23.44	2024.09	–	340.91	1.55	1.55
840725	1	36.446	27.12	2033.53	–	395.46	–	–
840725	10	36.442	26.65	2030.83	–	384.45	1.57	1.57
840816	1	36.454	28.09	2022.67	–	389.75	1.57	1.57
840816	10	36.411	27.29	2022.00	2369.34	391.95	1.58	1.59
840911	1	36.497	26.69	2018.15	–	360.14	1.64	1.64
840911	10	36.494	26.60	2017.38	–	357.97	1.62	1.62
841018	1	36.378	23.14	2020.00	–	325.71	1.61	1.62
841018	10	36.378	23.49	2021.61	–	332.44	1.57	1.57
841119	1	36.454	23.52	2023.94	2379.05	332.61	1.63	1.63
841119	10	36.456	22.44	2023.72	2379.99	317.81	1.64	1.63
841217	1	36.434	20.79	2030.08	2380.60	305.74	1.61	1.62
841217	10	36.431	20.76	2029.91	2379.38	306.44	1.61	1.62

TABLE A.1 (continued)

Date [YYMMDD]	Depth [m]	Sal	Temp [°C]	DIC [μmol kg⁻¹]	Alk [μmol kg⁻¹]	pCO$_2$ [ppm]	Orig ^{13}δ [per mil]	Actual ^{13}δ [per mil]
850205	1	36.489	19.12	2048.76	–	309.33	1.41	1.41
850205	10	36.508	19.21	2048.31	2383.18	309.35	1.46	1.46
850212	1	36.513	19.04	2051.09	2383.44	310.89	1.42	1.41
850212	10	36.509	18.96	2051.10	2385.21	307.88	1.46	1.46
850315	1	36.519	19.18	2049.63	2383.54	310.51	1.47	1.47
850315	10	36.520	19.03	2048.76	2384.35	306.54	1.47	1.47
850411	1	36.468	19.14	2046.47	2377.63	311.85	1.45	1.45
850411	10	36.460	19.11	2046.70	2375.84	310.36	1.46	1.46
850506	1	36.472	20.12	2046.33	2375.01	327.21	1.53	1.53
850625	1	36.483	26.13	2036.77	–	384.03	1.55	1.56
850625	10	36.475	25.76	2035.62	–	377.54	1.56	1.56
850722	1	36.433	27.23	2026.99	2374.84	392.39	1.61	1.61
850722	10	36.417	26.70	2026.25	–	379.37	1.56	1.56
850826	1	36.472	27.86	2019.59	–	380.00	1.62	1.62
850826	10	36.456	27.06	2017.37	–	366.77	1.64	1.63
851014	1	36.349	25.76	2004.60	–	337.87	1.70	1.70
851014	10	36.349	25.70	2004.98	2374.21	336.84	1.70	1.70
851107	1	36.398	24.81	2016.36	–	340.02	1.67	1.66
851107	10	36.398	24.89	2015.56	2376.36	340.22	1.63	1.63
851205	1	36.542	22.62	2026.00	2384.64	318.74	1.68	1.68
851205	10	36.541	22.57	2026.37	–	317.11	1.65	1.65
860203	1	36.551	19.97	2039.02	2387.65	301.31	1.62	1.62
860203	10	36.545	20.13	2039.59	2386.53	305.16	1.58	1.58
860405	1	36.434	20.10	2040.62	–	313.65	1.51	1.51
860405	10	36.440	20.00	2038.80	–	307.21	1.48	1.48

TABLE A.1 (continued)

Date [YYMMDD]	Depth [m]	Sal	Temp [°C]	DIC [µmol kg⁻¹]	Alk [µmol kg⁻¹]	pCO$_2$ [ppm]	Orig $^{13}\delta$ [per mil]	Actual $^{13}\delta$ [per mil]
860428	1	36.535	20.02	2042.08	–	308.14	–	1.54
860428	10	36.535	20.35	2041.11	–	310.84	–	1.48
860529	2	36.583	21.42	2041.81	2388.60	322.16	1.55	1.55
860529	12	36.580	21.77	2041.02	2390.01	323.78	1.55	1.55
860707	1	36.323	25.99	2021.28	2375.65	363.64	1.57	1.57
860707	10	36.371	26.00	2024.41	2374.84	370.33	1.60	1.60
860826	1	36.384	27.90	2014.99	2376.21	379.02	–	1.67
860826	10	36.378	27.74	2015.03	2376.88	375.99	–	1.65
860923	1	36.213	26.31	1999.82	2364.99	346.47	1.66	1.66
860923	10	36.207	25.95	2000.19	2364.22	343.32	1.63	1.63
861105	1	36.549	24.28	2018.55	2386.59	326.48	1.70	1.70
861105	10	36.555	24.07	2018.34	2387.09	323.10	1.68	1.68
861121	1	36.530	23.49	2019.22	2382.48	322.16	1.59	1.59
861121	10	36.530	23.48	2020.62	2380.70	326.02	1.59	1.59
870122	1	36.514	20.58	2040.53	2386.35	311.90	1.47	1.47
870122	10	36.509	20.47	2040.36	2387.06	309.49	1.46	1.47
870307	1	36.623	19.28	2053.76	2393.08	307.66	1.44	1.44
870307	10	36.621	19.26	2053.85	2393.83	306.69	1.42	1.42
870328	1	36.570	19.24	2054.52	2388.81	312.59	1.40	1.40
870328	10	36.565	18.87	2054.16	2389.41	306.78	1.40	1.40
870422	1	36.642	20.27	2051.40	2393.76	316.03	1.48	1.48
870422	10	36.640	20.18	2049.97	2392.95	313.79	1.47	1.47
870526	1	36.498	22.51	2035.02	2383.76	330.96	–	1.53
870526	10	36.524	22.13	2037.59	2384.38	329.40	–	1.62
870626	1	36.504	25.59	2028.34	2384.77	359.68	1.58	1.58
870626	10	36.500	24.96	2029.21	2385.23	352.11	1.57	1.57

TABLE A.1 (continued)

Date [YYMMDD]	Depth [m]	Sal	Temp [°C]	DIC [μmol kg⁻¹]	Alk [μmol kg⁻¹]	pCO$_2$ [ppm]	Orig [13]δ [per mil]	Actual [13]δ [per mil]
870718	1	36.529	27.19	2025.53	2386.02	375.33	1.52	1.52
870718	10	36.520	27.19	2026.25	2385.55	377.04	1.52	1.52
870827	1	36.454	25.95	2020.60	2378.21	359.98	1.56	1.56
870827	10	36.449	25.70	2020.06	2378.82	355.03	1.58	1.58
871019	1	35.959	26.03	1983.72	2338.16	348.40	1.57	1.57
871019	10	36.215	26.04	1997.21	2353.41	353.05	1.58	1.58
871111	1	36.428	24.33	2015.74	2376.56	333.44	1.56	1.56
871111	10	36.424	24.27	2016.07	2376.09	333.68	1.61	1.61
871203	1	36.514	22.14	2027.80	2384.51	315.31	1.52	1.50
871203	10	36.510	22.11	2027.44	2383.92	315.06	1.55	1.53
880125	1	36.492	19.79	2042.97	2383.76	308.29	1.35	1.33
880125	10	36.493	19.78	2042.70	2381.14	310.71	1.36	1.34
880309	1	36.538	19.09	2050.78	2386.44	307.90	1.34	1.32
880309	10	36.535	18.99	2050.33	2385.69	306.85	1.36	1.34
880512	1	36.532	20.58	2041.72	2376.24	325.32	1.41	1.39
880512	10	36.519	20.24	2041.29	2373.40	323.59	1.40	1.38
880613	1	36.512	23.29	2040.03	2384.97	347.30	1.34	1.45
880613	10	36.506	23.27	2039.52	2384.39	346.91	1.35	1.43
880721	1	36.279	26.93	2020.78	2370.89	381.43	1.34	1.46
880721	10	36.281	26.29	2021.25	2370.56	373.85	1.33	1.42
880824	1	36.360	28.26	2009.36	2373.61	377.77	–	–
880824	10	36.356	28.07	2009.06	2374.52	373.51	1.58	1.51
880920	1	36.357	27.88	2001.05	2372.69	360.52	1.64	1.57
880920	10	36.355	27.81	1999.60	2373.42	356.46	1.59	1.62
881117	1	36.382	23.52	2017.23	2379.20	322.16	–	1.55
881117	10	36.384	23.43	2017.19	2377.50	322.92	–	1.44

TABLE A.1 (continued)

Date [YYMMDD]	Depth [m]	Sal	Temp [°C]	DIC [μmol kg^{-1}]	Alk [μmol kg^{-1}]	pCO$_2$ [ppm]	Orig $^{13}\delta$ [per mil]	Actual $^{13}\delta$ [per mil]
890129	1	36.646	20.79	2040.11	2394.71	305.79	1.38	1.45
890129	18	36.645	20.79	2039.85	2394.77	305.37	1.40	1.48
890426	1	36.619	20.54	2050.44	2389.74	322.47	1.35	1.36
890426	10	36.611	20.40	2048.94	2390.62	317.48	1.40	1.39
890522	1	36.573	22.69	2042.60	2377.30	353.49	–	1.36
890522	10	36.580	22.34	2040.98	2376.85	346.86	–	1.38
890721	1	36.613	27.35	2036.94	2390.44	391.64	–	1.51
890721	10	36.600	26.86	2037.12	2393.61	380.70	–	1.54
890901	1	36.428	27.08	2018.75	2381.25	368.01	–	1.57
890901	10	36.457	26.93	2020.72	2385.19	364.47	–	1.57

75

TABLE A.2

Mixed-layer depth and vertical diffusion coefficient

Date [YYMMDD]	Mixed-layer depth [m]	Kz [10^4 m² s⁻¹]	Date [YYMMDD]	Mixed-layer depth [m]	Kz [10^4 m² s⁻¹]
830127	80	1.375	840127	85	0.705
830201	80	0.910	840210	130	2.508
830223	150	1.756	840309	150	1.458
830307	100	—	840403	120	3.845
830317	150	1.270	840410	170	2.235
830331	100	1.980	840430	130	1.849
830412	110	4.010	840627	28	0.182
830509	25	0.453	840628	28	0.182
830608	15	0.452	840710	25	0.109
830614	15	0.331	840724	15	0.104
830623	25	0.125	840808	15	0.136
830708	15	0.233	840816	15	0.071
830721	25	0.107	840829	15	0.200
830802	15	0.119	840911	25	0.106
830819	25	0.087	841003	25	0.105
830912	20	0.110	841018	50	0.108
830926	20	0.143	841102	75	0.240
831010	25	0.071	841117	100	0.347
831024	35	0.084	841130	80	0.325
831114	50	0.227	841217	100	0.700
831123	70	0.228			
831208	75	0.156			
831228	80	0.339			

TABLE A.2 (continued)

Date [YYMMDD]	Mixed-layer depth [m]	Kz [10^4 m^2 s^{-1}]	Date [YYMMDD]	Mixed-layer depth [m]	Kz [10^4 m^2 s^{-1}]
870122	135	0.965	880115	150	0.836
870219	270	1.448	880125	145	1.229
870307	175	3.269	880209	170	1.072
870328	220	–	880223	130	2.315
870416	55	1.536	880309	170	3.661
870422	30	0.871	880325	165	1.918
870507	15	0.522	880404	185	2.877
870526	25	0.209	880503	55	1.331
870608	18	0.522	880523	15	–
870626	20	0.484	880613	15	0.099
870707	15	0.251	880711	15	0.116
870718	15	0.078	880721	15	0.126
870817	25	0.093	880815	15	0.095
870817	25	0.093	880824	25	0.071
870827	25	0.101	880920	25	0.125
870831	15	0.101	881026	75	0.161
871019	30	0.200	881117	90	0.094
871026	50	0.121	881130	100	0.334
871111	50	0.156	881228	105	0.525
871116	80	0.191			
871124	50	0.352			
871127	65	0.290			

TABLE A.2 (continued)

Date [YYMMDD]	Mixed-layer depth [m]	Kz [10^4 m^2 s^{-1}]	Date [YYMMDD]	Mixed-layer depth [m]	Kz [10^4 m^2 s^{-1}]
870122	135	0.965	880115	150	0.836
870219	270	1.448	880125	145	1.229
870307	175	3.269	880209	170	1.072
870328	220	–	880223	130	2.315
870416	55	1.536	880309	170	3.661
870422	30	0.871	880325	165	1.918
870507	15	0.522	880404	185	2.877
870526	25	0.209	880503	55	1.331
870608	18	0.522	880523	15	–
870626	20	0.484	880613	15	0.099
870707	15	0.251	880711	15	0.116
870718	15	0.078	880721	15	0.126
870817	25	0.093	880815	15	0.095
870817	25	0.093	880824	25	0.071
870827	25	0.101	880920	25	0.125
870831	15	0.101	881026	75	0.161
871019	30	0.200	881117	90	0.094
871026	50	0.121	881130	100	0.334
871111	50	0.156	881228	105	0.525
871116	80	0.191			
871124	50	0.352			
871127	65	0.290			

TABLE A.2 (continued)

Date [YYMMDD]	Mixed-layer depth [m]	Kz [10^4 m^2 s^{-1}]
890108	110	0.762
890118	100	0.609
890210	200	2.906
890228	150	–
890315	60	1.384
890504	20	0.609
890522	15	0.408
890613	15	0.085
890627	15	0.138
890721	15	0.069
890731	20	0.060
890901	25	0.092
890915	25	0.132

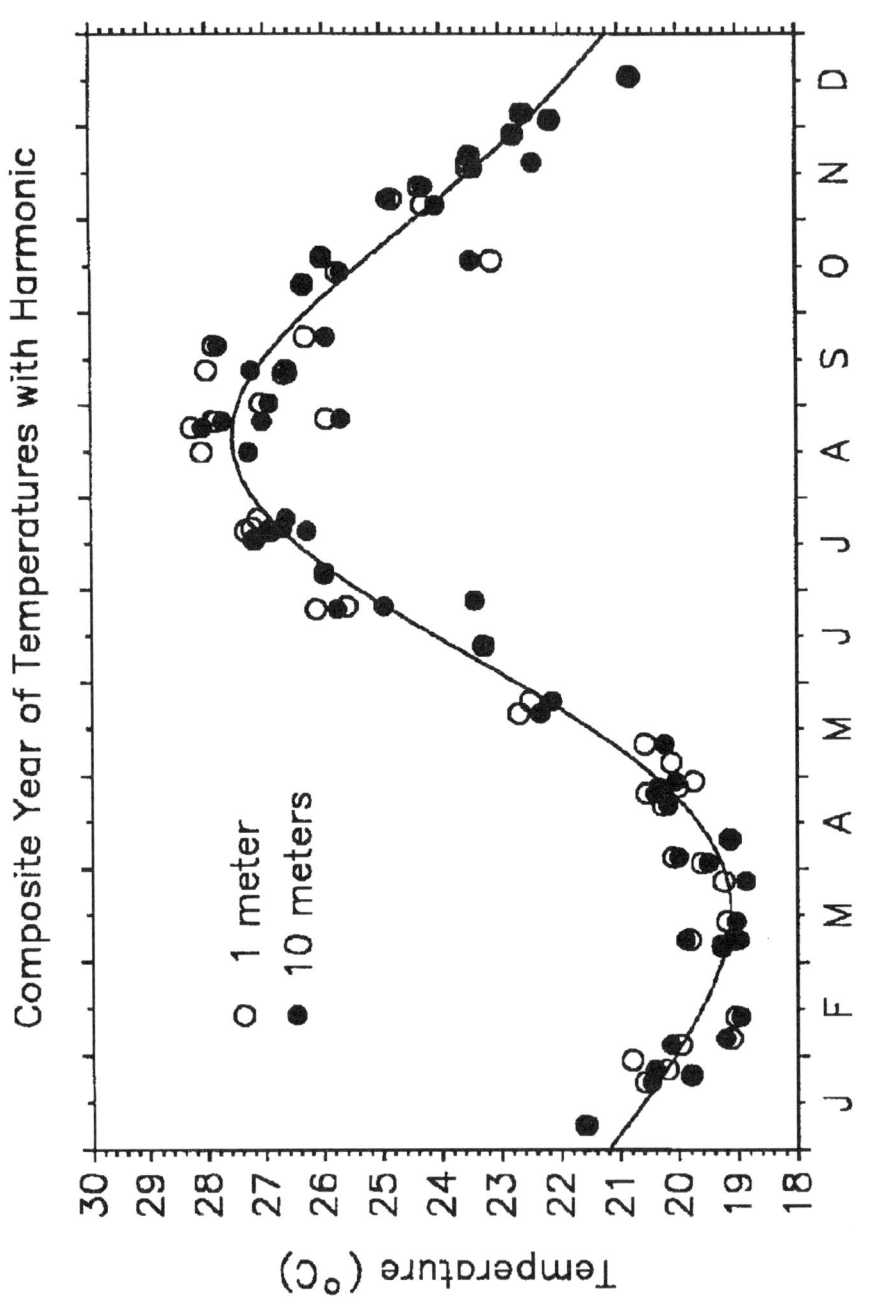

Figure 1. Seasonal variation in ocean water temperature of the surface mixed layer at Station S, in °C. Observations from 1983 to 1989, at 1 and 10 m, are shown as open and solid circles, respectively. The smooth curve depicts a two-harmonic fit through the data (see text and Table 1).

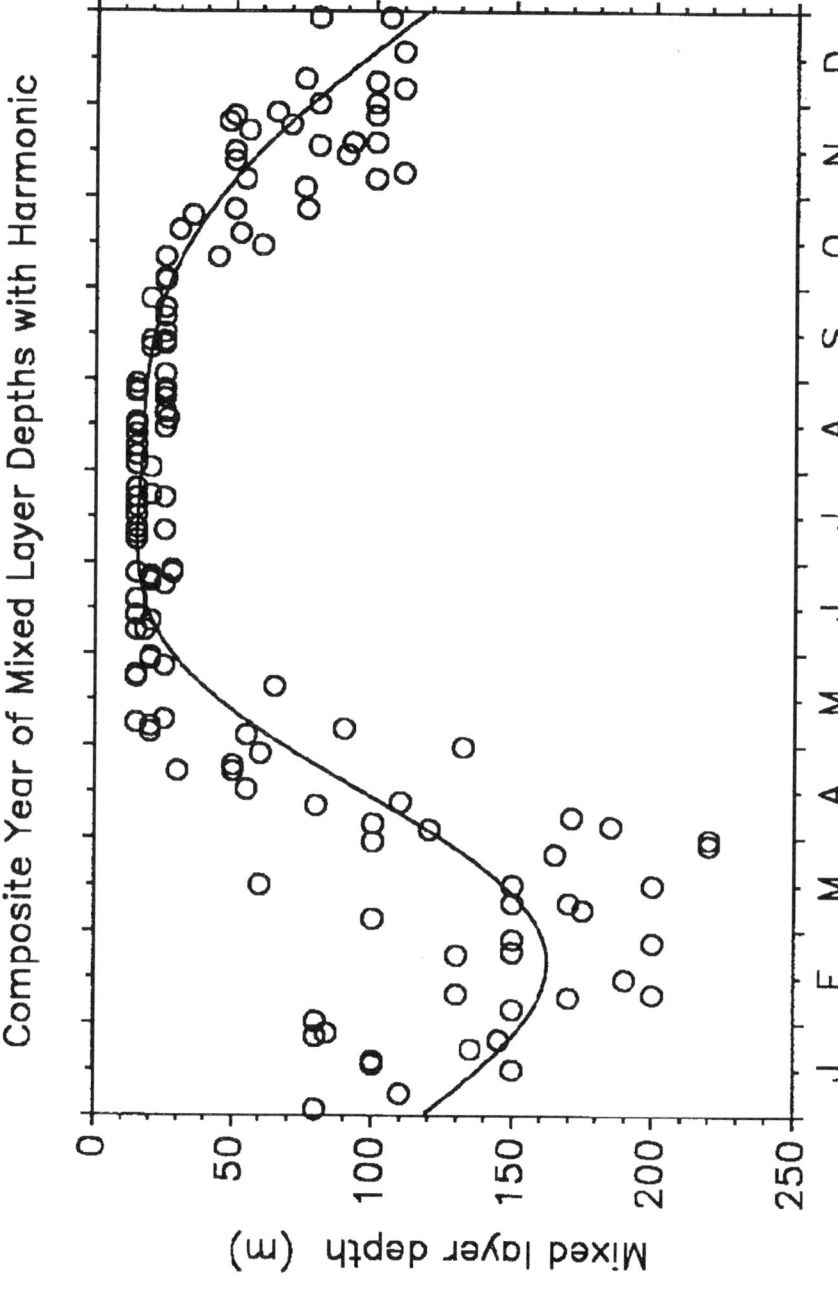

Figure 2. Seasonal variation in depth of the surface mixed layer at Station S, in meters. As described in the text, the depth was determined by the criterion of Levitus (1982) from vertical profiles of density calculated from observations of temperature and salinity for the same period as the data of Figure 1. The smooth curve is a three-harmonic fit to the depth estimates. On the basis of this curve the mixed layer shoals from day 51 through day 190 of the annual cycle.

81

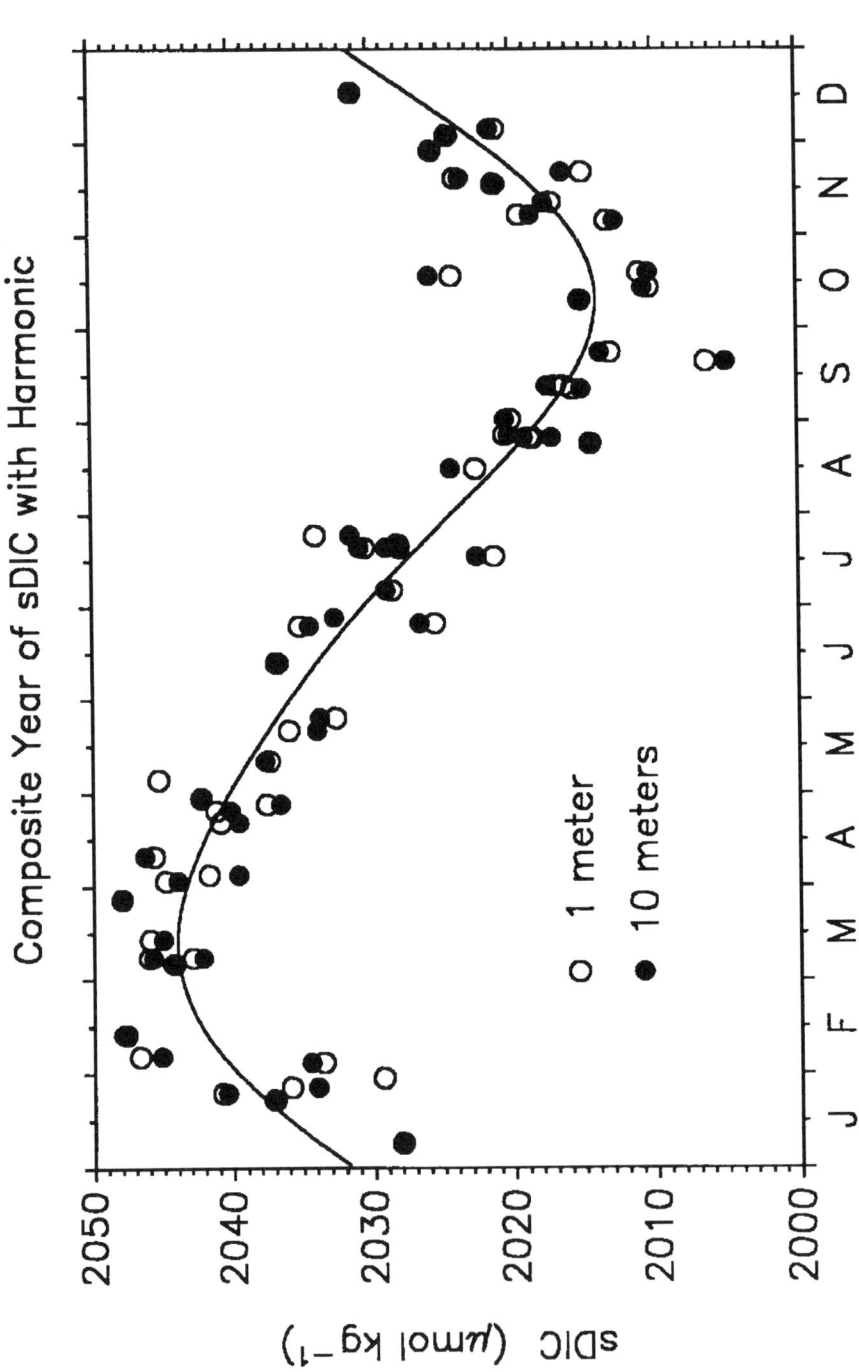

Figure 3. Seasonal variation in dissolved inorganic carbon (sDIC), in μmol kg⁻¹, in the surface mixed layer at Station S. The data, normalized to the average salinity of 36.452, are based on observations from 1983 to 1989 as in the previous figures. The smooth curve is a two-harmonic fit through the data. As in Figure 1, observations were taken at 1 and 10 m and are shown by open and closed circles, respectively.

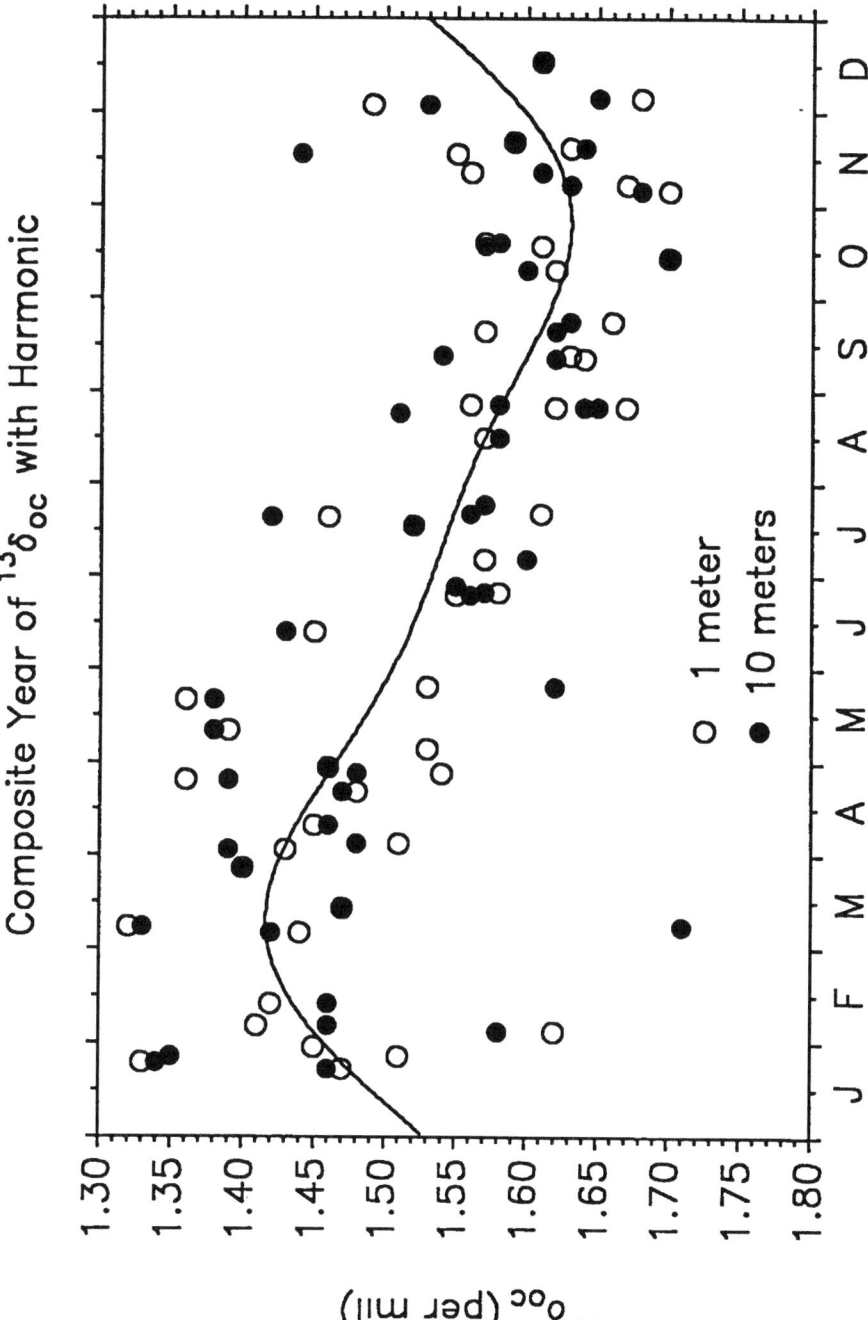

Figure 4. Seasonal variation in the reduced $^{13}C/^{12}C$ isotopic ratio of DIC, $^{13}\delta_{oc}$, in per mil, in the surface mixed layer at Station S. The symbols and curve depict data and a harmonic fit as in Figure 3. The vertical axis is inverted so that addition of carbon of organic origin to the DIC pool results in higher values on the plots of both concentration and isotopic ratio.

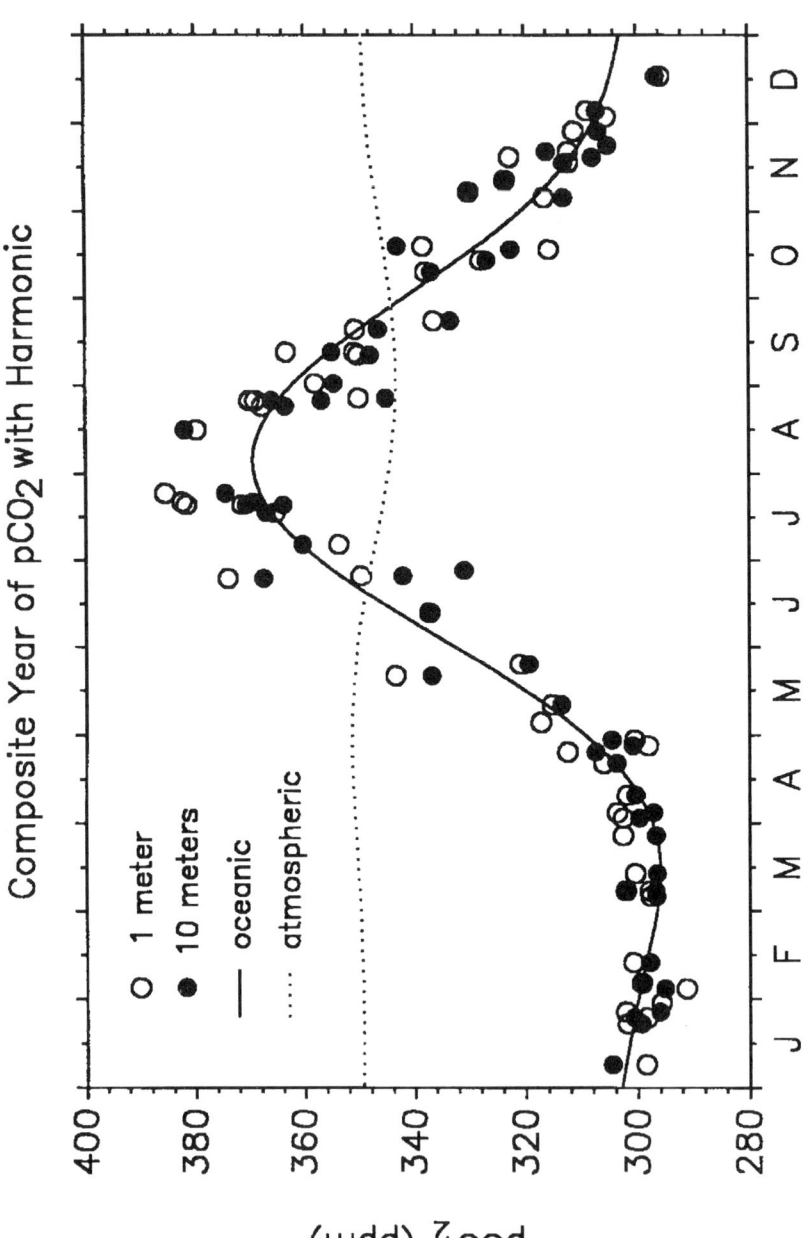

Figure 5. Seasonal variation in the partial pressure of CO_2 at the sea-surface, pCO_{2oc}, determined from DIC and alkalinity measurements, as described in the text. The symbols and solid curve depict data and a harmonic fit as in Figure 3. A dashed curve depicts an additional harmonic function representing the seasonal variation in atmospheric CO_2 partial pressure, pCO_{2am}, derived from observations at La Jolla, California. Both air and seawater pCO_2 values are expressed as a mixing ratio in parts per million of dry air (ppm, approximately equal to μatm).

84

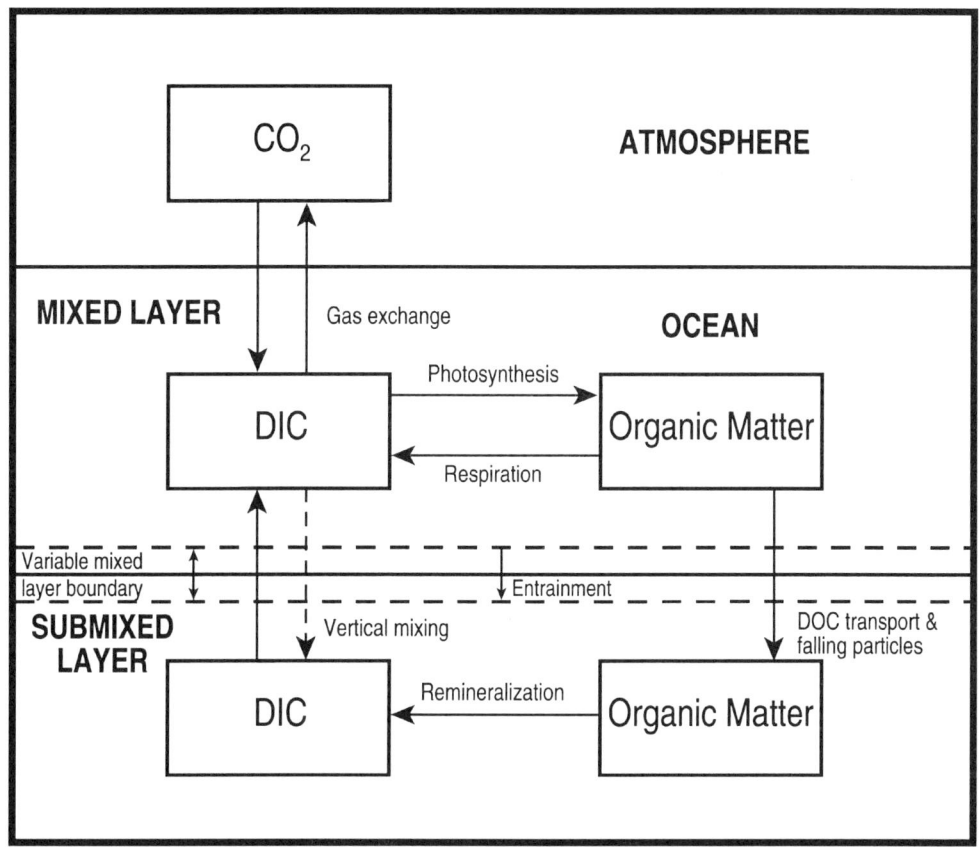

Figure 6. Diagram of the pathways and reservoirs of the carbon cycle in the surface mixed layer near Bermuda, as employed in the seasonal model. The model neglects photosynthesis and respiration below the mixed layer. These fluxes are assumed to cancel with a portion of the remineralization flux, which is thus also neglected.

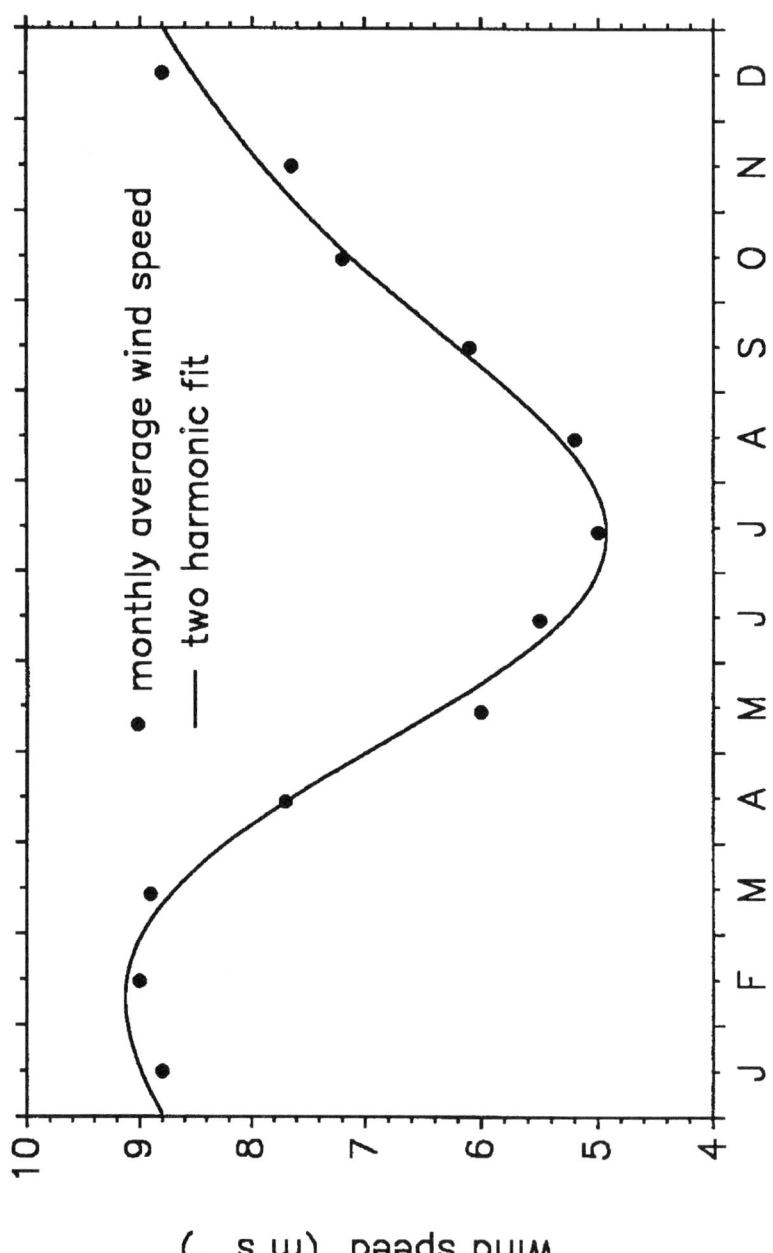

Figure 7. Seasonal variation in wind speed, in m s⁻¹, at Station S defined by monthly average values shown as dots [data from Musgrave et al. (1988)]. The smooth curve is a two-harmonic fit through the monthly data. The measurements are in reference to a height of 10 m above the sea-surface.

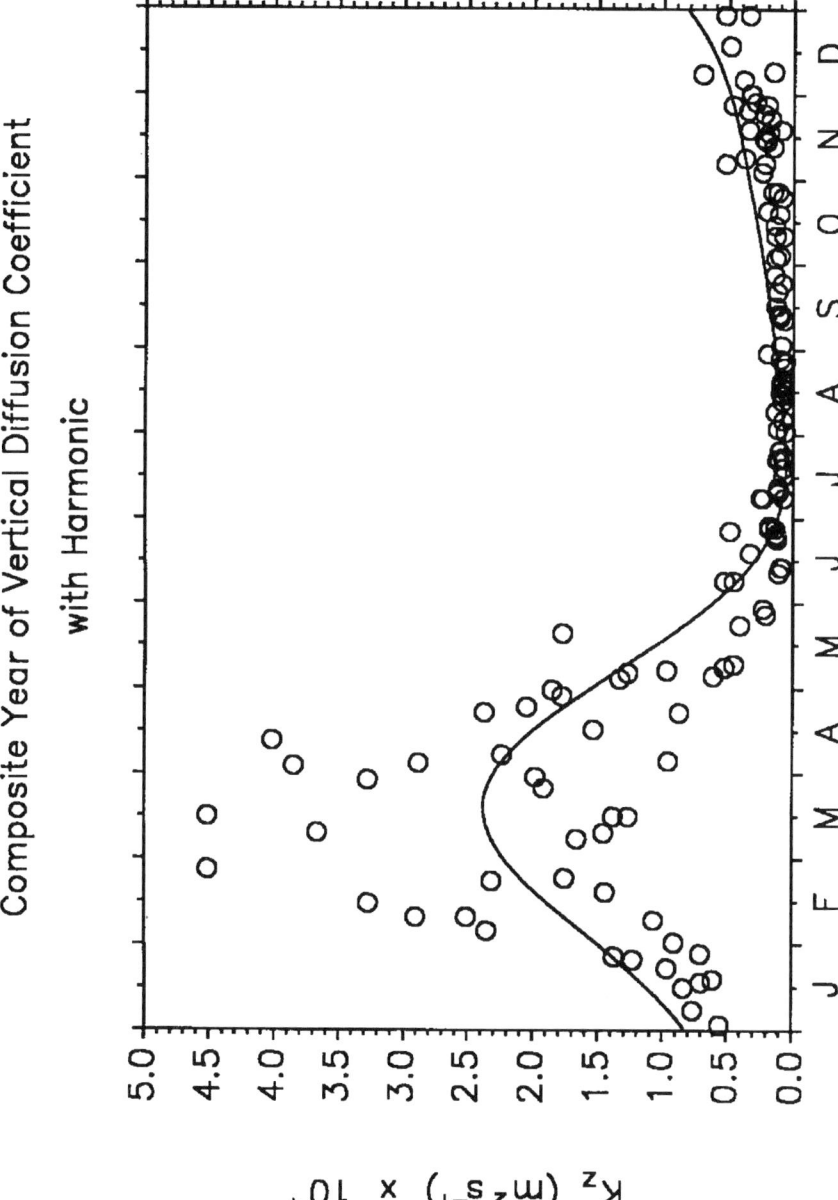

Figure 8. Seasonal variation in the vertical diffusion coefficient at the base of the surface mixed layer, in $m^2 s^{-1}$, computed from the buoyancy frequency as described in the text. Individual values of the coefficient, estimated from observations of salinity and temperature for the same period as in Figure 1, are shown as open circles. The smooth curve is a two-harmonic fit through the data.

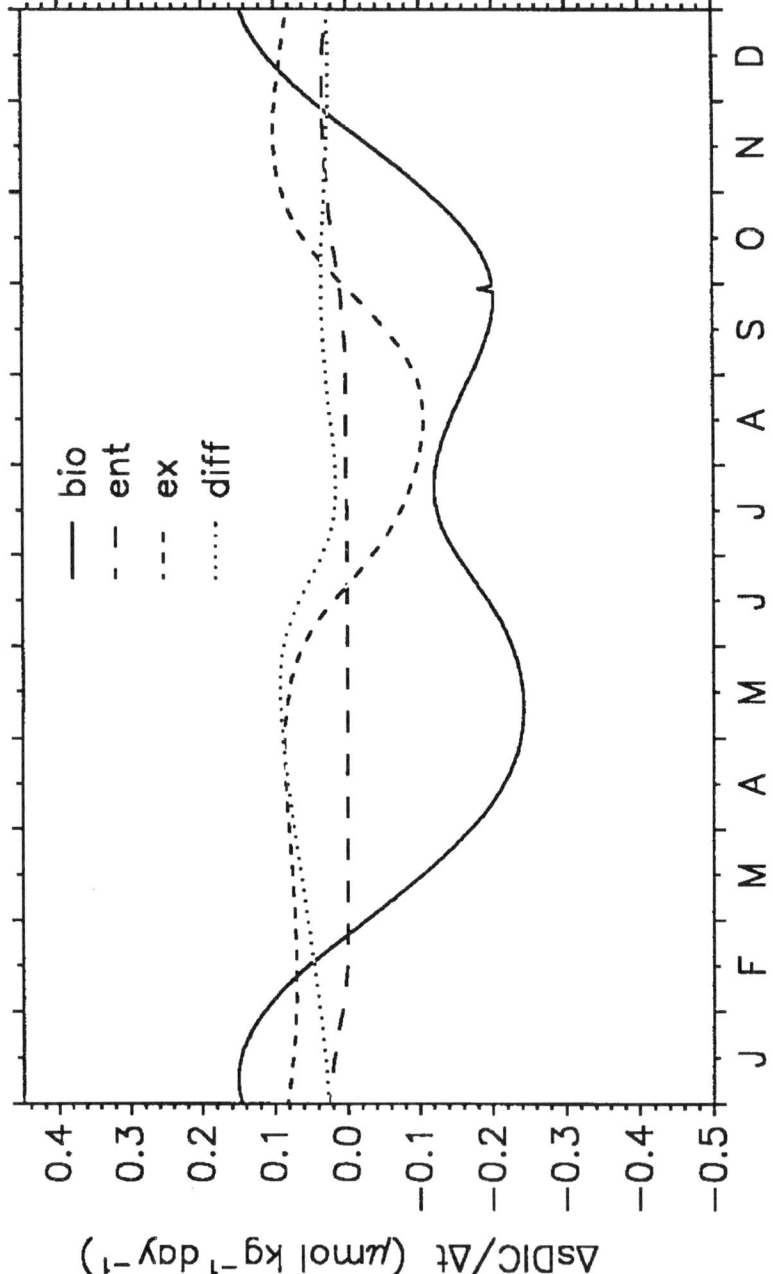

Figure 9. Computed seasonal variations in the rate of change in sDIC, in μmol kg^{-1} day^{-1}, produced by net biological exchange, net air-sea gas exchange, vertical diffusive transport, and vertical entrainment. Each variation is shown by a separate curve, as labeled.

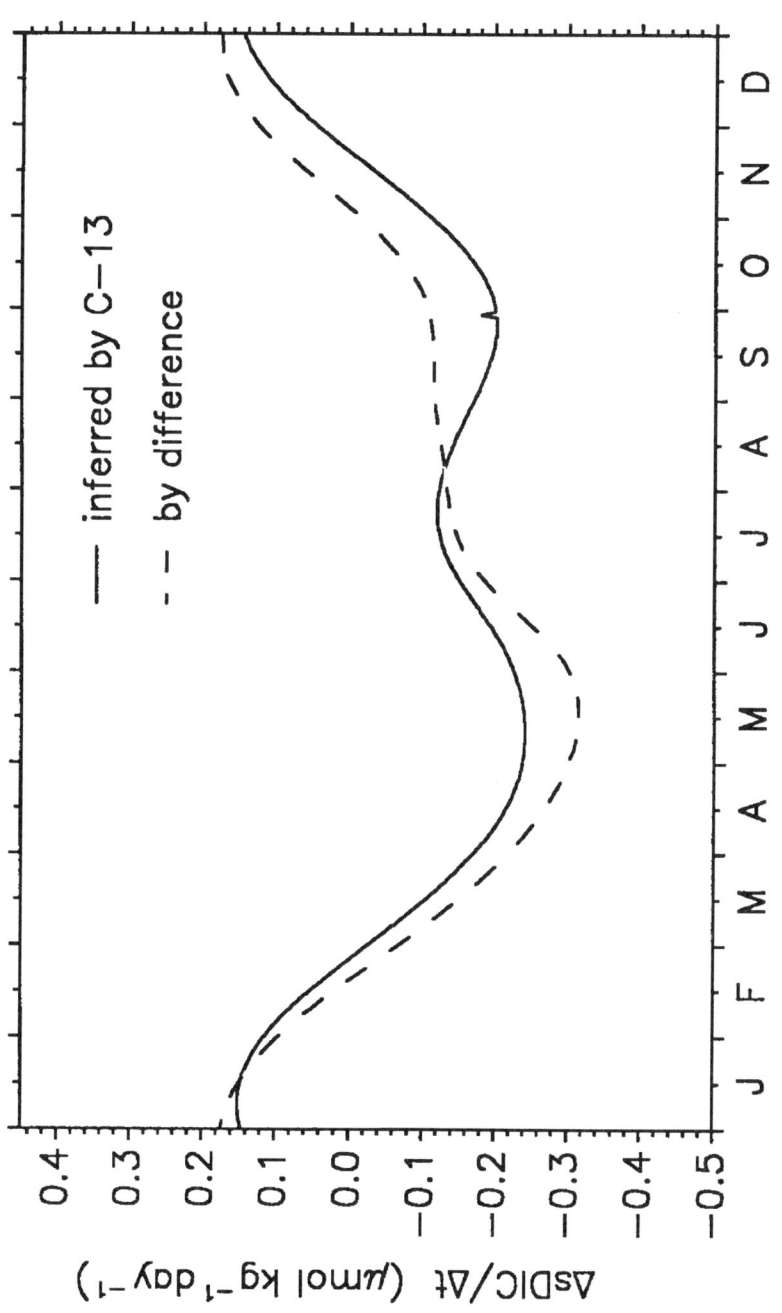

Figure 10. Computed seasonal variations in the rate of change in sDIC, in μmol kg⁻¹ day⁻¹, produced by net biological exchange calculated two ways. The solid line, identical to the solid line in Figure 9, was calculated by observing changes in ¹³C/¹²C isotopic ratio of DIC. The dashed curve was calculated from the differences in sDIC fluxes without regard to isotopic changes.

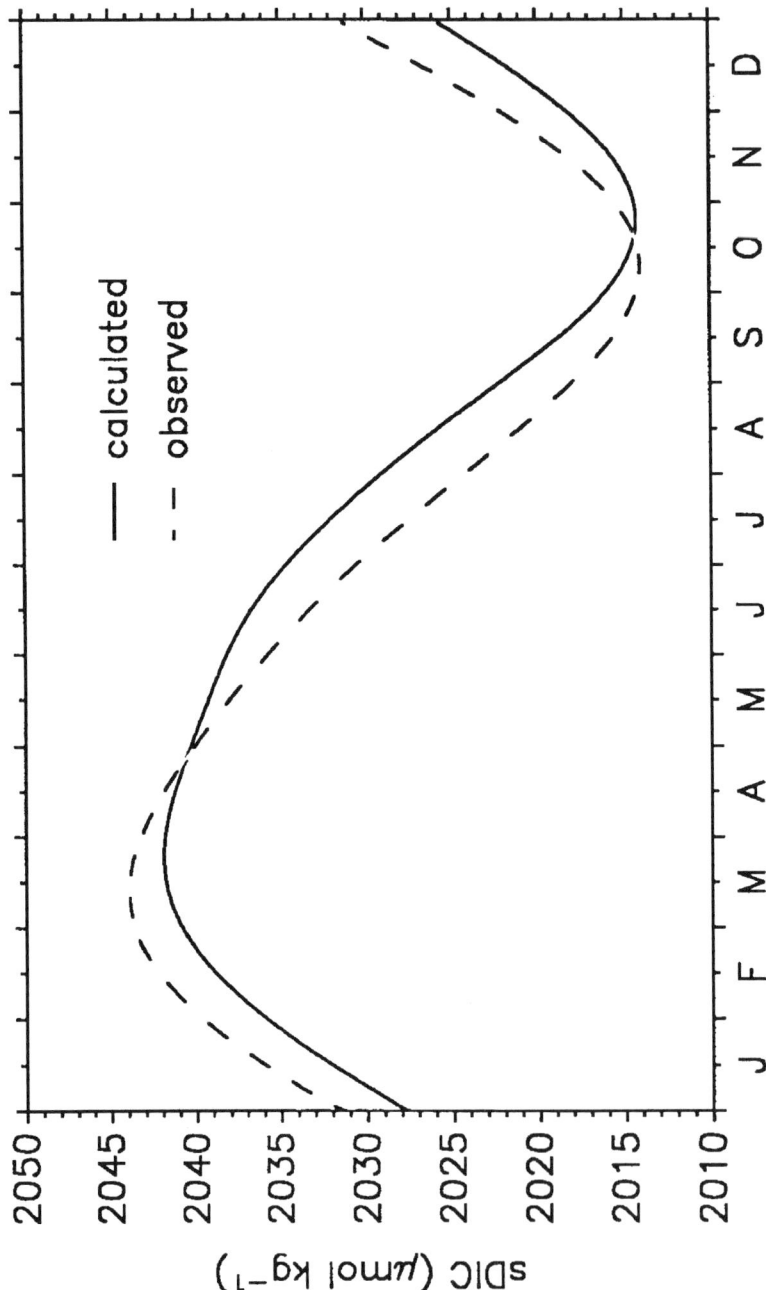

Figure 11. Seasonal variation in sDIC, in μmol kg^{-1}, computed as the cumulative sum of daily increments plotted in Figure 10. The observed variation is shown for comparison.

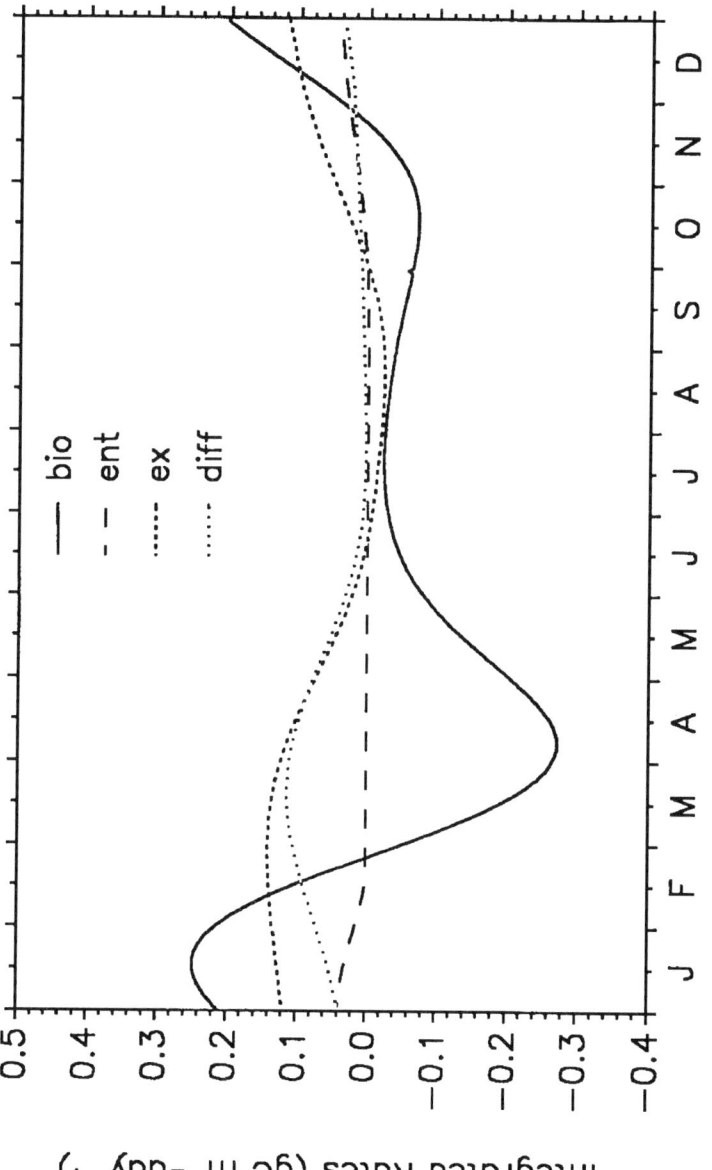

Figure 12. Computed seasonal variations in the rates of change in sDIC, vertically integrated over the surface mixed layer, in gC m^{-2} day^{-1}, for the four processes shown in Figure 9.

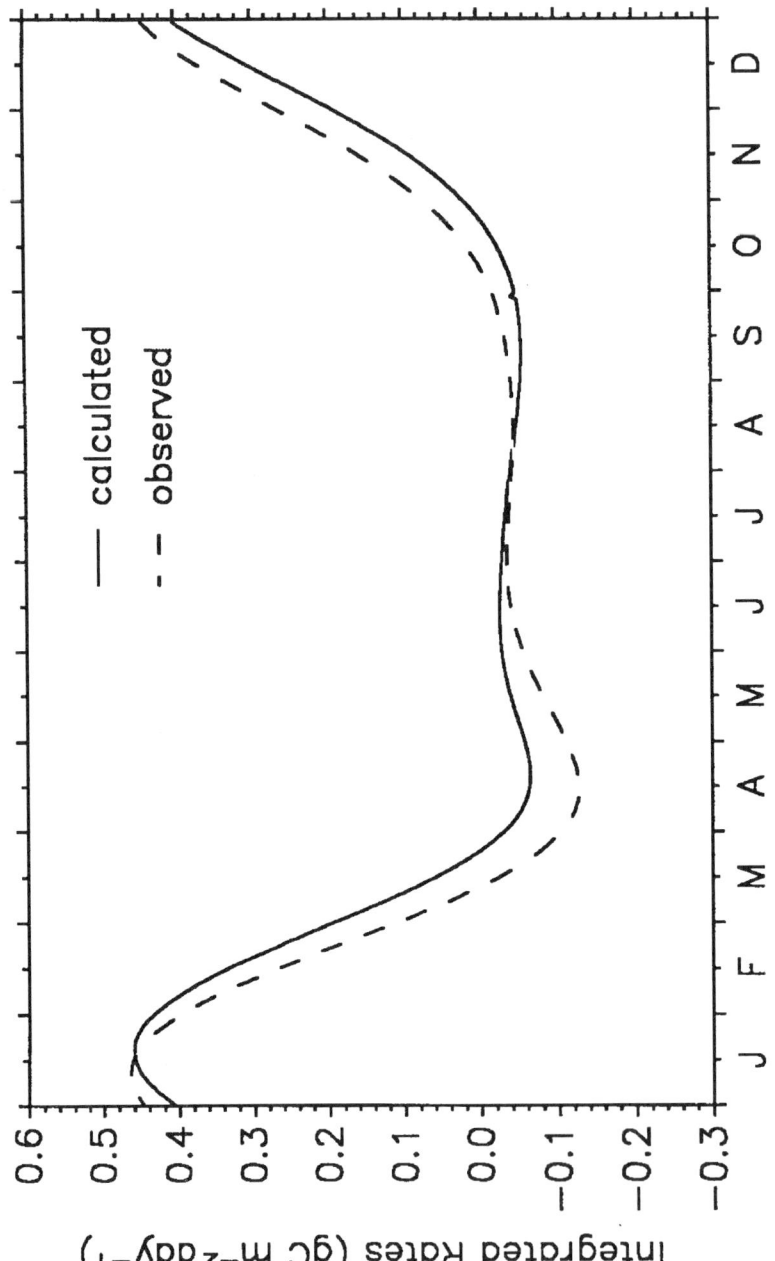

Figure 13. Seasonal variation in the rate of change in sDIC vertically integrated, in gC m^{-2} day^{-1}, computed as the cumulative sum of the four processes shown separately in Figure 12. The observed rate of change is shown for comparison.

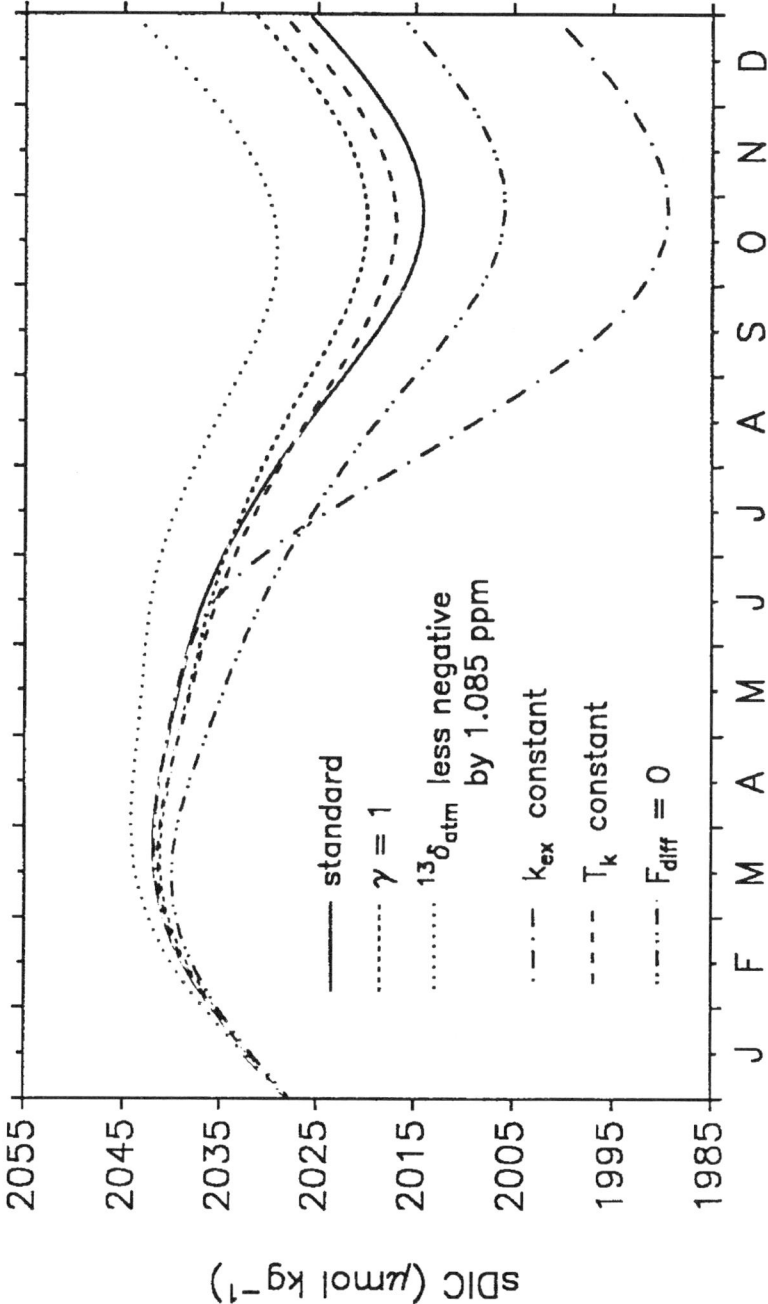

Figure 14. Seasonal variation in sDIC, as computed for various sensitivity tests described in the text. The standard case (solid curve), shown previously in Figure 11, is shown again for comparison.

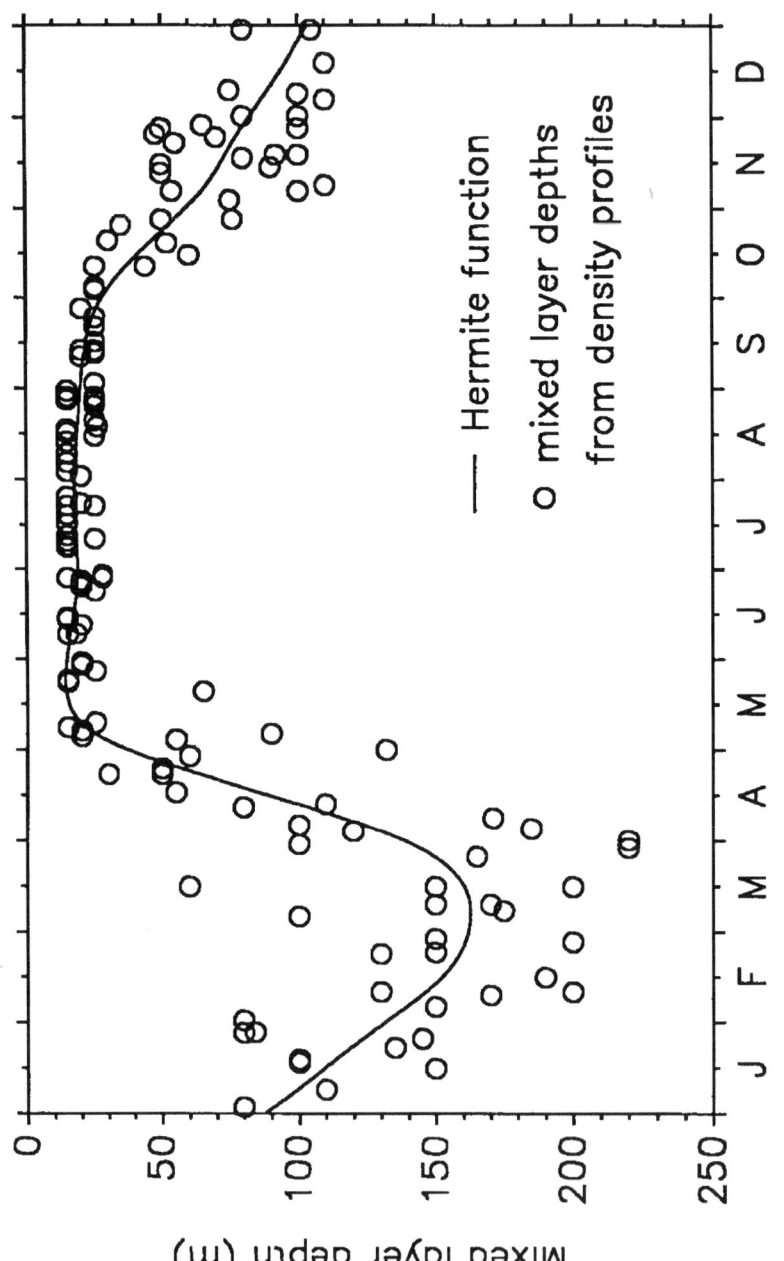

Figure 15. Seasonal variation in the depth of the surface mixed layer, in meters, as in Figure 2, except that the seasonal cycle, represented by the smooth curve, is derived from a fit to Hermite polynomials instead of by means of a harmonic function.

— Hermite function

○ mixed layer depths from density profiles